THE ATLAS OF
FOOD

WHO EATS WHAT, WHERE AND WHY

Erik Millstone and Tim Lang

Consultant:
Axel Drescher

Earthscan Publications Ltd, London

First published in the UK in 2003
by Earthscan Publications Ltd

A catalogue record for this book is available from
the British Library

ISBN: 1 85383 965 5

Produced for Earthscan Publications Ltd by
Myriad Editions Limited
6–7 Old Steine, Brighton BN1 1EJ, UK
http://www.MyriadEditions.com

Edited and co-ordinated for Myriad Editions by
Jannet King and Candida Lacey
Design and graphics by Paul Jeremy,
Isabelle Lewis, Corinne Pearlman
Maps created by Isabelle Lewis
Additional research by Jannet King

Printed and bound in Hong Kong
Produced by Phoenix Offset Limited
under the supervision of Bob Cassels,
The Hanway Press, London

For a full list of publications please contact:

Earthscan Publications Ltd
120 Pentonville Road, London, N1 9JN, UK
tel: +44 (0)20 7278 0433
fax: +44 (0)20 7278 1142
email: earthinfo@earthscan.co.uk
http://www.earthscan.co.uk

Earthscan is an editorially independent subsidiary of
Kogan Page Ltd and publishes in association with
WWF-UK and the International Institute for
Environment and Development

THE ATLAS OF
FOOD

Erik Millstone is a Reader in Science Policy
at the University of Sussex, UK.
He has been working on food-related issues since
author of *Food Additives*;
s: A Guide for Everyone;
Lead and Public Health,
al and magazine articles
on the politics of food.

Tim Lang is Professor of Food Policy
at City University, London. He is chair of Sustain and
a consultant to the World Health Organization and
other governmental and non-governmental bodies.
He works on food policy and the public interest,
linking public and environmental health with
consumers and social justice.
He is the co-author of *The Unmanageable
Consumer*; *The New Protectionism*
and the forthcoming *Food Wars*.

Axel Drescher is Professor in Applied Geography
at Freiburg University.

In the same series:

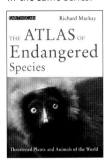

THE ATLAS OF ENDANGERED SPECIES
by Richard Mackay

By the same authors:

LEAD AND PUBLIC HEALTH
by Erik Millstone

THE NEW PROTECTIONISM
by Tim Lang and Colin Hines

FOOD WARS
by Tim Lang and Michael Heasman
(forthcoming)

CONTENTS

CONTRIBUTORS

The authors would like to thank the following people for their contributions on specific topics:

Peter Backman (Foodservice Intelligence, London)
38 Eating Out

David Barling (Thames Valley University, London)
21 Agricultural Biodiversity

Paul Brassley (Seale-Hayne Faculty, University of Plymouth, UK) 10 Mechanization

David Buffin (Pesticide Action Network UK)
17 Pesticides

Colin Butler (National Centre for Epidemiology & Population Health, Australian National University, Canberra) 3 Environmental Challenges; 4 Water Pressure

Dominique Charron (Centre for Infectious Diseases, Prevention and Control, Health Canada)
5 Consuming Disease

Alizon K Draper (Centre for Food, Nutrition and Public Health, School of Integrated Health, University of Westminster, London)
6 Under-nutrition; 31 Staple Foods; 32 Changing Diets

Axel Drescher (University of Freiburg, Germany)
19 Urban Farming

Joan Gandy (Centre for Food, Nutrition and Public Health, School of Integrated Health, University of Westminster, London) 6 Under-nutrition; 31 Staple Foods; 32 Changing Diets

Michael Heasman (Visiting Research Fellow, Centre for Food Policy, Thames Valley University, London) 35 Functional Foods

Sarah Herd (Department of Population Medicine, University of Guelph, Ontario, Canada)
5 Consuming Disease

Andy Jones (Denbighshire, UK)
23 Trade Flows; 26 Food Miles; 29 Developing Trade

Tim Lobstein (The Food Commission, London)
41 Advertising

Vivien Lund (London)
8 Food Aid; 9 Food Aid as Power

Mary-Ann McKibben (Alcohol Concern)
40 Alcohol Consumption

Sue Mayer (GeneWatch, UK)
15 Genetic Modification; 16 Genetically Modified Crops

Geof Rayner (Chair, UK Public Health Association Visiting Research Fellow, Centre for Food Policy, Thames Valley University) 40 Alcohol Consumption

Mike Rayner (Department of Public Health, University of Oxford) 7 Over-nutrition

Peter Stevenson (Compassion in World Farming, UK)
13 Industrial Farming; 25 Animal Transport in Europe

Steve Suppan (Institute for Agriculture and Trade Policy, Minneapolis, USA) 28 Trade Disputes

Geoff Tansey (Hebden Bridge, UK)
14 Agricultural R&D

David Waltner-Toews (Department of Population Medicine, University of Guelph, Ontario, Canada)
5 Consuming Disease

INTRODUCTION

WHAT WE EAT, where we eat and how we eat reveals a world of food and drink culture. How our daily bread – or rice – reaches our plates and palates is sometimes so complex that we cannot unravel its route in one bite. That is why we wanted to provide a series of snapshots of the key features of the modern food system.

We all, whether growers or consumers, sellers or cooks, need to know more about the food we eat – why we get to eat when some don't, and what trends and technologies characterize our food supply.

The role of food in western societies has changed; and it is changing in other areas of the world too, as trade and politics, technologies and patterns of demand restructure economies. Everywhere our relationship with food is altered by innovatory food processing technologies and "fast food". Just think of all the gadgets made for kitchens in the 20th century. Even the meaning of cooking is changing. For many it is no longer a routine task, but a creative weekend activity. It has moved from the ghetto of women's magazines to soft travel features and TV entertainment. Ironically, TV cooking and food programs mushroom almost in inverse proportion to the amount of cooking that actually occurs.

The changes in the humble act of cooking symbolize wider changes in much to do with food. Some see this as the result of globalization, others as the pursuit of rationality. Whatever the reasons, this atlas explores how everything to do with food has experienced an unprecedented period of flux: on the farm, in the factory, on retail shelves, in transit, in marketing and the home. As a result, the health patterns associated with food and diet have also changed. Under-nutrition and over-consumption coexist in both the industrialized and the developing world.

The world of food also offers a mixture of sobering and energizing insights into humanity's relationship with the natural world. On the one hand there are the astonishing leaps in total production, which have confounded the gloomy predictions of half a century ago that the world's population would outgrow its capacity to produce enough food to go round. On the other hand, it seems that no sooner has a policy mountain been climbed, than more peaks appear. For every plant-breeding "breakthrough" or technological innovation, the harsh reality is that the world of food is still characterized by localized shortages and surpluses, plant, animal and human disease, price fluctuations and poor quality produce. In compiling the atlas, we have sought to highlight these changes.

Most food originates on farms or from fisheries, and is passed along a supply chain that includes traders, processors, transporters and retailers until it eventually reaches us as consumers. But this chain is far from simple, and involves a web of conflicting interests and complex relationships between the agents involved. Farmers are dependent on industrial sectors to provide them, for example, with seeds, fertilizers, pesticides, machinery and fuel. The choices they make are often influenced by what happens in local and global markets, on the financial trading floors, in factories, laboratories and banks. The kind of food that is produced, and the methods used to produce it, are increasingly influenced by those further down the food chain – the processors and the retailers. And they, in turn, claim to defer to consumers. Yet while consumers are supposedly the arbiters of modern taste, as individuals we often feel powerless to influence it. We inhabit a food world in which the trader and retailer are often more powerful than either the primary producer or end consumer.

The international food trade provides a further set of interlocking links in the food chain. Increasing amounts of the food we eat comes from abroad. Yet chronic hunger is to be found in countries where agricultural exports are expanding. While economists stress the importance of developing countries earning foreign currency through export trade, there is no nutritional logic to the process by which food is transferred from areas where under-consumption is a serious problem to areas where the dominant diet-related problems are over-consumption. Often food is exported from countries where under-nutrition is endemic. Just as Ireland exported food during its Great Famine of the 1840s, so Africa exports to Europe today.

The industrialization of the European and US agricultural and food systems, now spread worldwide, is today strongly influenced by the subsidies that have been used to try to enhance the economic viability of domestic industries. Subsidies lead, almost inevitably, to over-production, and food surpluses are disposed of in ways that create further loops and links in the food chain. A large proportion of the agricultural crops produced in the industrialized countries is not consumed directly by people, but indirectly in the form of meat, since vast quantities of grains and pulses are used as animal fodder. This can be economically efficient, but nutritionally and ecologically highly inefficient. Other surplus food is exported, often with the benefit of further subsidies. The net effect is to drive down the price of internationally traded food commodities, which in turn makes it harder for farmers in developing countries to compete or thrive.

A fundamental theme of the atlas is that even though poor nutrition afflicts an estimated one third of the world's population (2 billion out of 6 billion), this is not a consequence of any overall scarcity of food, but more often of poverty, lack of resources and access to food. More food is produced on this planet than would be needed to feed everyone adequately, but political, economic, environmental and social forces result in inequitable production, distribution and consumption. The world needs food justice.

Some people assume that hunger is a technical problem that will be solved by the rapid introduction of new technologies. But caution is needed when transferring technology from one society to another. Crop varieties and production methods that require high inputs (fertilizers, pesticides and mechanical harvesting) tend to benefit relatively wealthy farmers but disadvantage poorer farmers, who are unable to make the necessary investment. While technological change may result in more food being produced overall, it may increase the number of hungry farmers and amplify the inequalities between rich and poor. A change to less labor-intensive agricultural practices can have a devastating effect in areas where the vast majority of people are employed on the land.

In the industrialized world, the food system has, in modern times, been characterized by high rates of technological change and sustained growth in productivity. Technological innovations continue, but their introduction and diffusion are becoming more, rather than less, problematic. While US farmers and consumers have largely accepted the introduction of genetically modified crops, their European and Japanese counterparts have been reluctant to do so. And while "functional foods" that promise positive health benefits may be at the cutting edge of industrial innovation in the food industry, their adoption and diffusion will probably be rather uneven.

As the interconnections and loops of the food-supply chain have become increasingly complex, the distance between producers and consumers has widened – both literally and metaphorically. One effect has been rising consumer concern and mistrust. The strong, if relatively short-lived, public rejection of beef in the UK in 1996, when it was admitted that BSE-contaminated meat was responsible for the degenerative brain condition vCJD, underlined how vulnerable industrialized farming is to volatile public opinion. The food corporations recognize this, which is one reason why they pour vast sums into advertising and sponsorship – in order to create trust based on brand loyalty.

A more tangible outcome of the distance between producer and consumer is the enduring problem of microbiological food poisoning that continues to afflict poor and rich communities alike. As the food chain has lengthened, and increased in complexity, the opportunities for pathogens to spread has increased too. Barriers to the transfer of contamination have been undermined by the astonishing speed and distance traveled by both food and people – all fueled by cheap oil. Without plentiful oil, shipping, flying and trucking foods immense distances would be inconceivable. Evidence of the impact of fuel emissions and their role in climate change suggests that this practice may be unsustainable.

The direction of change in the global food system is not, however, one-way. Worldwide, a movement of informed groups and alliances has emerged, questioning current practices, demanding change, making connections. The Fairtrade movement, for example, is attempting to redress the balance in favor of the small producer, while other international movements represent the interests of laborers and consumers against the power of multinational corporations and those with powerful vested interests. Some are collecting information, others campaigning and implementing alternatives. The increase in certified organic farms, and the importance of urban farming demonstrates a growing demand for shorter food chains, and a far more direct relationship between producer and consumer.

This atlas provides evidence of the growing recognition that the modes of production, processing, distribution and consumption that prevail – because, in the short-to-medium-term they are the most profitable – are not necessarily the most healthy or the most environmentally sustainable. It is unlikely that the agricultural and food system can remain as it is. Given its inherent instability, the food supply chain is bound to change. It changed 10,000 years ago, with settled agriculture, and again with enclosures and plant and animal breeding. Later, the internal combustion engine transformed factories and shops. Today, the food supply chain is challenged by a combination of biotechnologies and huge external forces such as climate change, migration and urbanization. The outcome of these pressures will in part depend on political decisions yet to be taken. Our motivation for the atlas is to provide the basis for an understanding of such underlying issues and to argue for more rational and more just food policies.

The story of contemporary food is a tale of how human societies are fractured by a combination of "givens" and "takens" – some eat, others starve; some toil, others consume; some profit, others pay; some live, others die; some see the big picture of food, others deny it. We hope you find the information in these pages helps you to respond to the fundamental food challenges that face us all.

We would like to thank all those who helped us to produce this book. The contributors on specific topics are listed on page 6, but other people helped point us in the right direction, among them Martin Caraher, Barry Leathwood, Tony McMichael, Terry Marsden, Mark Ritchie, Patti Rundall, Bill Vorley and Kevin Watkins. The information and PR desks of Fairtrade Labelling Organisation and a number of research and trade bodies, including Burger King, IGD, KFC and PGD, also provided statistics. We would like to thank Axel Drescher for his input, and acknowledge the backup given by Yannick Borin and Sylvie Fritche of the Center for Food Policy.

The simplicity of the spreads belies the time, care and attention to detail that went into them. Above all, we thank Jannet King and the team at Myriad Editions – Paul Jeremy, Candida Lacey, Isabelle Lewis and Corinne Pearlman – for creating the maps and graphics, and for keeping us on track.

Finally, we would like to thank our families for their patience and support.

ERIK MILLSTONE and TIM LANG
Summer 2002

CONTEMPORARY CHALLENGES

1

"Food is no longer viewed first and foremost as a sustainer of life. Rather, to those who seek to command our food supply it has become instead a major source of corporate cash flow, economic leverage, a form of currency, a tool of international politics, an instrument of power – a weapon!"
– A V Krebs, *The Corporate Reapers: The Book of Agribusiness*, 1992

1 FEEDING THE WORLD

Over- and under-nutrition

Prevalence of underweight and obese adults
1999 percentages

 underweight

overweight

8.2%
5.8%

worldwide

8.9%

1.8%

least developed countries

6.9%
4.8%

developing countries

17.1%

economies in transition

2.4%

20.4%

developed market economies

1.6%

TWO BILLION PEOPLE suffer from chronic under-nutrition, and 18 million die each year from hunger-related diseases. And yet more than enough food is produced to feed everyone in the world.

If all the cereal produced each year was divided evenly among the world's population, everyone would receive more than is needed for them to survive. Instead, in many African, Asian and Latin American countries the average calorie intake per person is substantially below that needed to remain healthy, and malnutrition is rife, especially among children. In most industrialized countries the opposite is true: the average intake of calories is significantly *above* the 2,500 calories recommended as healthy by nutritionists, leading to a different kind of malnutrition – over-nutrition.

Averages hide wide disparities, however, and everywhere there are extremes of under- and over-nutrition. In countries whose economies are developing fast, such as Brazil, China, India and a "nutrition transition" is taking place. While their poorest citizens still suffer from under-nutrition, their more prosperous citizens are switching to western-style diets, high in fats and sugar, leading to increased obesity and the health problems it brings.

There are large surpluses of food – in particular grain – on the global food market. Some of the food grown in the wealthier countries is destroyed to protect the market price. Huge quantities of grains and pulses harvested from North American fields are fed to cattle, pigs, and poultry. This conversion of grain into meat for human consumption wastes much of the nutritional value of the grain.

Food-rich countries send large quantities of their surplus food to countries where it is in short supply. But even where this redistribution of food is well-handled at the local level, it does not offer a long-term solution. The long-term reduction of hunger and malnutrition worldwide requires an increase in income levels brought about by improvements in local food production and better access to food.

40 million

people die
of hunger
each year
while

356

kilograms
of grain
per person
is being
produced

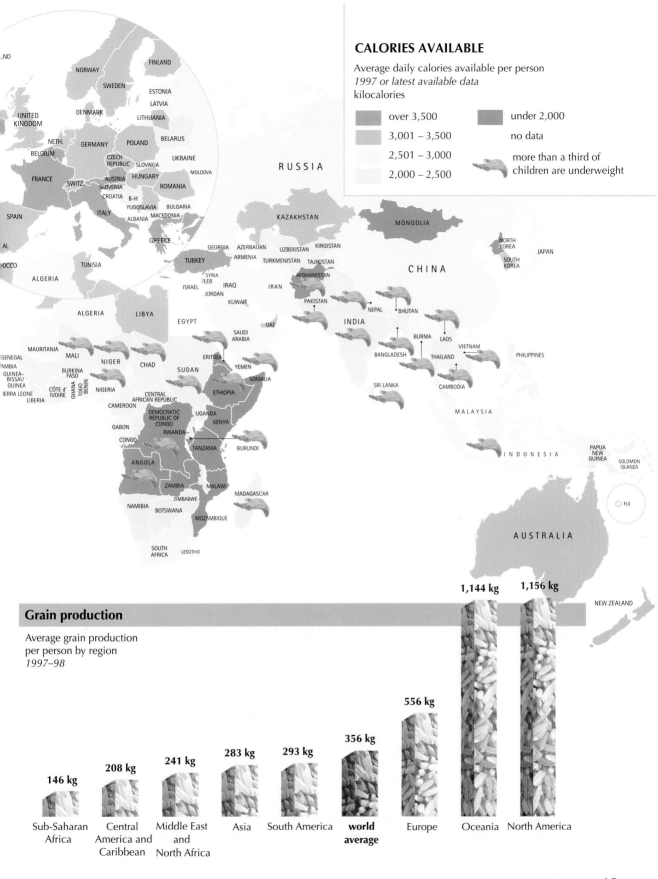

CALORIES AVAILABLE

Average daily calories available per person
1997 or latest available data
kilocalories

- over 3,500
- 3,001 – 3,500
- 2,501 – 3,000
- 2,000 – 2,500
- under 2,000
- no data
- more than a third of children are underweight

Map labels

ND, NORWAY, FINLAND, SWEDEN, ESTONIA, LATVIA, DENMARK, LITHUANIA, UNITED KINGDOM, NETH., GERMANY, POLAND, BELARUS, BELGIUM, CZECH REPUBLIC, SLOVAKIA, UKRAINE, FRANCE, SWITZ., AUSTRIA, HUNGARY, MOLDOVA, SLOVENIA, ROMANIA, CROATIA, B-H, SPAIN, ITALY, YUGOSLAVIA, BULGARIA, ALBANIA, MACEDONIA, AL, GREECE, RUSSIA

KAZAKHSTAN, MONGOLIA, NORTH KOREA, JAPAN, SOUTH KOREA, GEORGIA, AZERBAIJAN, KIRGISTAN, ARMENIA, TURKMENISTAN, UZBEKISTAN, TAJIKISTAN, CHINA, OCCO, TUNISIA, TURKEY, SYRIA, LEB, ISRAEL, IRAQ, IRAN, AFGHANISTAN, JORDAN, KUWAIT, PAKISTAN, NEPAL, BHUTAN, ALGERIA, LIBYA, EGYPT, SAUDI ARABIA, UAE, INDIA, BURMA, LAOS, VIETNAM, PHILIPPINES, BANGLADESH, THAILAND

MAURITANIA, MALI, NIGER, CHAD, ERITREA, YEMEN, SENEGAL, AMBIA, GUINEA-BISSAU, GUINEA, BURKINA FASO, SUDAN, SOMALIA, SRI LANKA, CAMBODIA, CÔTE d'IVOIRE, GHANA, TOGO, BENIN, NIGERIA, ETHIOPIA, IERRA LEONE, LIBERIA, CENTRAL AFRICAN REPUBLIC, CAMEROON, UGANDA, KENYA, MALAYSIA, GABON, DEMOCRATIC REPUBLIC OF CONGO, RWANDA, CONGO, TANZANIA, BURUNDI, INDONESIA, PAPUA NEW GUINEA, SOLOMON ISLANDS, ANGOLA, ZAMBIA, MALAWI, MADAGASCAR, FIJI, ZIMBABWE, NAMIBIA, BOTSWANA, MOZAMBIQUE, AUSTRALIA, SOUTH AFRICA, LESOTHO, NEW ZEALAND

Grain production

Average grain production per person by region
1997–98

- Sub-Saharan Africa — **146 kg**
- Central America and Caribbean — **208 kg**
- Middle East and North Africa — **241 kg**
- Asia — **283 kg**
- South America — **293 kg**
- **world average** — **356 kg**
- Europe — **556 kg**
- Oceania — **1,144 kg**
- North America — **1,156 kg**

13

2 POPULATION AND PRODUCTIVITY

THE AMOUNT OF FOOD produced in many countries is insufficient to feed the population, even at current levels. The challenge is to shift productivity from those countries currently producing an excess, to those where people are undernourished.

The world's population is unevenly distributed, and a map of the world in which the countries are sized according to the number of people that inhabit them, shows Asia and Southeast Asia looming large by comparison with North America. The rate at which populations are increasing also varies widely, with countries in the tropics and the southern hemisphere increasing more rapidly than those in the north. (Indeed, in parts of Europe, the population is actually falling.)

During the second half of the 20th century, not only did the world's population grow but its rate of expansion accelerated. Predictions for further growth vary, but even if the rate of increase slows down, the world's population is still likely to exceed 9 billion by 2050, with more than half of all people living in cities. The countries currently unable to produce sufficient food to keep people healthy are also the ones facing the most rapid population increase. Although food is redistributed around the globe, either as trade or aid – or a mixture of both – such redistribution is neither sufficient to solve the problem of under-nutrition, nor desirable as a long-term solution.

Improvements are needed in agricultural practices and in social structures so that more food can be produced and consumed where it is most needed. The amount of food produced on each hectare of agricultural land varies widely between regions, and is roughly in an inverse relation to the pattern of need created by the rate of population increase. And although productivity improved markedly in South America and Asia in the last quarter of the 20th century, in Africa, where the need is greatest, there was very little change.

POPULATION INCREASE

Annual percentage change
2000

- 3% and over
- 2% – 2.9%
- 1% – 1.9%
- 0% – 0.9%
- decrease
- no data

Countries' shares
of world population
2000 percentages

☐ = 1%
▫ = 0.1%

CANADA

IRELAND

USA

DOMINICAN REPUBLIC

CUBA

MEXICO   HAITI

GUATEMALA
EL SALVADOR   HONDURAS
                     NICARAGUA
COSTA RICA

                     VENEZUELA
COLOMBIA

ECUADOR   BRAZIL

PERU

BOLIVIA   PARAGUAY
CHILE   URUGUAY

ARGENTINA

Productivity

Regional cereal yield
1999
kilograms per hectare

- 5,000 kg and over
- 3,000 kg – 4,999 kg
- 2,000 kg – 2,999 kg
- 1,500 kg – 1,999 kg
- under 1,500 kg
- no data

↑ increase in productivity
1975–1999

↓ decrease in productivity
1975–1999

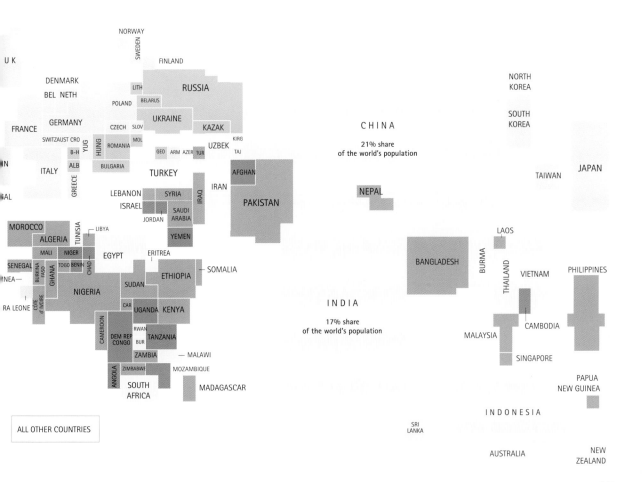

NORTH AMERICA 61%

WESTERN EUROPE 70%

EASTERN EUROPE 19%

ISRAEL 1%

ASIA 77%

JAPAN 1%

AFRICA 15%

SOUTH AMERICA 80%

SOUTH AFRICA 39%

OCEANIA 37%

NORWAY
SWEDEN
FINLAND
UK
DENMARK
BEL NETH
LITH
RUSSIA
POLAND
BELARUS
FRANCE
GERMANY
CZECH SLOV
UKRAINE
KAZAK
SWITZ AUST CRO
MOL
KIRG
UZBEK
TAJ
B-H YUG HUNG
ROMANIA
GEO ARM AZER TUR
ITALY
ALB
BULGARIA
TURKEY
AFGHAN
GREECE
LEBANON SYRIA
IRAQ
IRAN
PAKISTAN
ISRAEL
JORDAN SAUDI ARABIA
MOROCCO
TUNISIA
LIBYA
YEMEN
ALGERIA
MALI NIGER
EGYPT
ERITREA
SENEGAL
TOGO BENIN
CHAD
SOMALIA
NEA—
BURKINA FASO
GHANA
ETHIOPIA
RA LEONE
CÔTE d'IVOIRE
NIGERIA
SUDAN
CAR
UGANDA KENYA
CAMEROON
RWAN
TANZANIA
DEM REP CONGO
BUR
ZAMBIA
MALAWI
ANGOLA
ZIMBABWE
MOZAMBIQUE
SOUTH AFRICA
MADAGASCAR

NORTH KOREA
SOUTH KOREA
CHINA
21% share of the world's population
TAIWAN
JAPAN
NEPAL
LAOS
BANGLADESH
BURMA
THAILAND
VIETNAM
PHILIPPINES
INDIA
17% share of the world's population
CAMBODIA
MALAYSIA
SINGAPORE
PAPUA NEW GUINEA
INDONESIA
SRI LANKA
AUSTRALIA
NEW ZEALAND

ALL OTHER COUNTRIES

15

DESPITE IMPROVEMENTS in agricultural practices, major environmental challenges are likely to make it difficult even to maintain the present level of global productivity, let alone increase yields.

In many countries there is a shortage of good quality soil, and increasing demands are being placed on limited water supplies. Global warming is leading to less stable weather conditions, and is adversely affecting harvests.

In Central and South America, and in Southeast Asia, rainforests are cleared in an attempt to extend agricultural land. In the short term, this creates more crop and grazing land, but the soil is usually so poor that it rapidly degrades.

In some areas of Africa, although the soil and climate are suitable for agriculture, endemic sleeping sickness prevents it from being used.

The picture is not totally bleak. Some forms of land degradation can be reversed, while the rate of progression of others can be slowed.

New approaches, including grains that are genetically engineered to better cope with an excessively hot climate, may overcome some of these emerging problems. A new strategy, involving the release of millions of sterilized tsetse flies (which transmit sleeping sickness) also shows some promise. In Asia, where there is little room for expansion of the agricultural area, global warming may enable farmers to move higher up mountain slopes and to more northerly latitudes.

International cooperation is essential to reduce greenhouse gas emissions and encourage education and economic opportunities for people in poorer countries, to allow them to prepare for the disruptions and challenges that lie ahead.

Soil degradation

Degree of degradation affecting world's agricultural land
2001

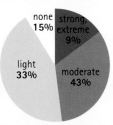

none 15%
strong, extreme 9%
light 33%
moderate 43%

Over 80 percent of arable land worldwide is affected by sufficient soil degradation to reduce its productivity.

Global harvests increased during the second half of the 20th century but they would have grown by a further 10 percent if it were not for cumulative soil degradation.

In addition to the constraints shown below, soil may also be too water logged or too slow to drain, or contain insufficient organic matter.

Some soil problems occur naturally; others result from human action, such as the excessive use of fertilizers, or over-working. Either way, they adversely affect the productivity of the soil.

Main types of soil degradation worldwide
2001

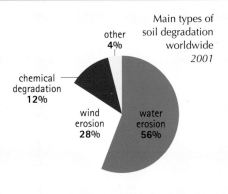

other 4%
chemical degradation 12%
wind erosion 28%
water erosion 56%

Major soil constraint
regional percentages
2001

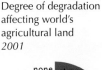

- no constraint
- high acidity
- high alkalinity
- low potassium
- high sodicity
- other constraints

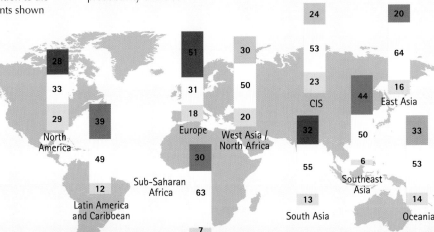

North America
28
33
29
39

Europe
51
31
18

West Asia / North Africa
30
50
20

Sub-Saharan Africa
49
30
63

Latin America and Caribbean
12
7

CIS
24
53
23

East Asia
20
64
16

Southeast Asia
44
50
32
6

South Asia
55
13

Oceania
33
53
14

Climate change is likely to affect global food security in several ways. A rise in the sea level, already occurring as a result of thermal expansion, threatens to inundate low-lying cropland in countries such as Bangladesh before the end of the century. And if saltwater were to intrude into underground aquifers in coastal areas around the world, it would make the land too saline for agriculture, as well as reducing the availability of drinking water. Higher average temperatures and drier winters are already reducing the size of glaciers in the Andes, the Himalayas and the Alps, affecting the flow of some of the world's major rivers.

Climate change also threatens surprises, such as a shutdown of the Gulf Stream that would paradoxically cause cooling in Europe, and reduce European food production. Climate change is already bringing unpredictable weather patterns, including intense tropical storms at unseasonable times, which damage crops.

Even if the amount of food that was produced under global warming were to remain the same by increased production at more northerly latitudes, it is possible that harvests in South Asia, Indonesia and Sub-Saharan Africa will decline, making them increasingly dependent on imported food, with serious political, economic and social consequences.

Climate change in 2100
Projected temperature changes if the atmospheric concentration of carbon dioxide doubles compared with 1996 levels

- more than 6°C increase
- 5°C – 6°C increase
- 4°C – 5°C increase
- 3°C – 4°C increase
- 2°C – 3°C increase
- 1°C – 2°C increase
- little or no change
- 0°C – 1°C decrease
- 1°C – 2°C decrease
- 2°C – 3°C decrease
- 3°C – 4°C decrease
- more than 4°C decrease

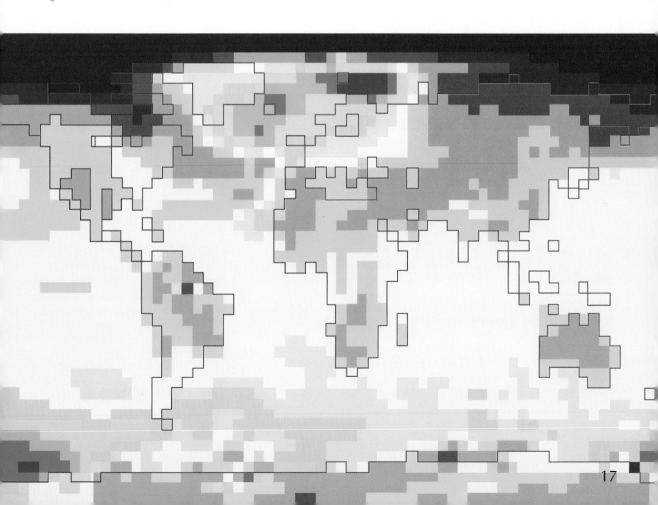

4 | WATER PRESSURE

Water shortfall

Predicted annual water shortfall in California by 2020

in a year of average rainfall — **782**

in a drought year — **2,019**

billion gallons

18%
of the world's cropland is irrigated and it produces

35%
of the world's harvest

MANY COUNTRIES already have insufficient fresh water. Many more are expected to experience water scarcity or water stress by 2050.

This map shows the average amount of water available per person. Water quality varies widely as does people's access to it, and the seasonal and geographic distribution. In the USA, for example, California's burgeoning population is putting an increasing strain on the state's limited resources. In China the wheat-growing north is far more water-stressed than its largely rice-growing south. The wheat harvest depends both on underground aquifers and on water from the Yellow River, which now fails to reach the sea for many days each year. Some countries, such as Egypt, are heavily dependent on water flowing in from another country and this increases their vulnerability.

Irrigated crops are crucial to food security, but for many countries and small farmers irrigation equipment is too expensive. Around 35 percent of the world's harvest is produced on the 18 percent of global cropland that is irrigated. In countries in the Middle East, irrigation is vital to agriculture. Over 20 percent of irrigated land worldwide has been affected by salinity, and even though the area of irrigated land is expanding by about 2 percent a year, a further 0.4 percent is being lost to salinity. The rate of expansion is also slowing, as accessible water becomes scarcer.

Some countries compensate for water-scarcity by importing grain and meat. In China, the scarcity in the agricultural north is aggravated by river water being diverted to more profitable industrial uses. While this may generate currency to pay for imported wheat to offset any shortfall, it makes China dependent on the global wheat market, and increases its food insecurity.

Many less industrialized countries, especially those in Africa, are even more vulnerable to water stress: when they experience drought they are too poor to buy food elsewhere.

CANADA

USA

MEXICO

CUBA

JAMAICA

BELIZE
HONDURAS

GUATEMALA

HAITI

DOMINICAN REPUBLIC

NICARAGUA

EL SALVADOR

COSTA RICA

PANAMA

VENEZUELA

GUYANA

SURINAME

COLOMBIA

ECUADOR

PERU

BRAZIL

BOLIVIA

PARAGUAY

CHILE

URUGUAY

ARGENTINA

Drought

African countries that experienced drought

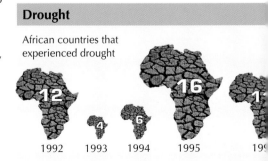

12 1992   **4** 1993   **6** 1994   **16** 1995   **1** 199

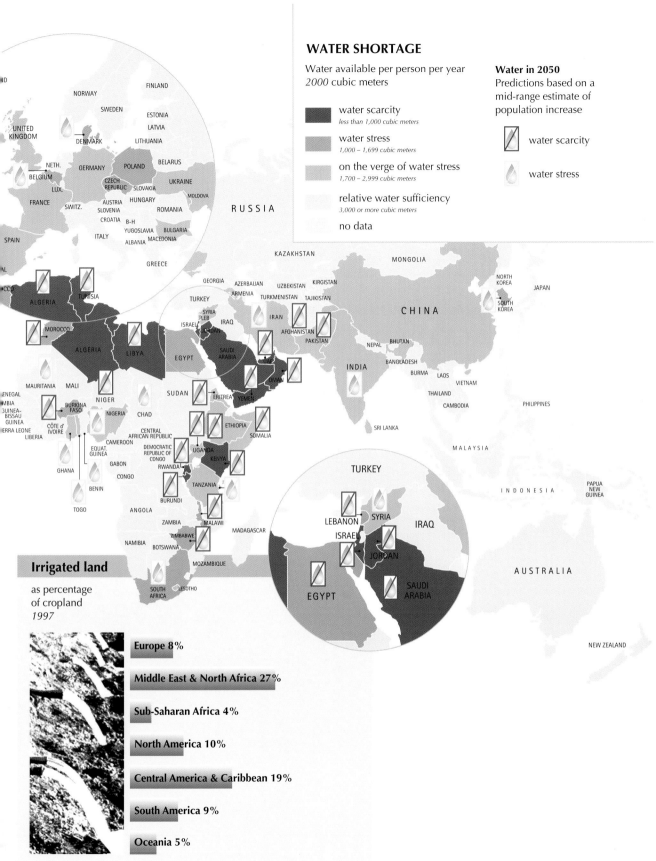

WATER SHORTAGE

Water available per person per year
2000 cubic meters

- **water scarcity**
  less than 1,000 cubic meters
- **water stress**
  1,000 – 1,699 cubic meters
- **on the verge of water stress**
  1,700 – 2,999 cubic meters
- **relative water sufficiency**
  3,000 or more cubic meters
- no data

Water in 2050
Predictions based on a
mid-range estimate of
population increase

- water scarcity
- water stress

Irrigated land

as percentage
of cropland
1997

Europe 8%

Middle East & North Africa 27%

Sub-Saharan Africa 4%

North America 10%

Central America & Caribbean 19%

South America 9%

Oceania 5%

CONSUMING DISEASE

Disease agents

- bacteria
- parasites
- toxins
- viruses and other agents

Europe

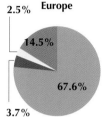

2.5%
14.5%
67.6%
3.7%

Central and South America

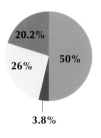

20.2%
50%
26%
3.8%

A BILLION PEOPLE each year suffer from diarrhea, most likely contracted from consuming contaminated food or water.

Around 4.5 million die from the condition. Most are in developing countries, but the problem is also increasing in the industrialized world. The globalization of food production and trade, and the popularity of foreign travel has led to food-borne diseases crossing borders and continents. With more people eating food prepared out of the home, the risk of infection is also increased.

Over 200 disease agents can be transmitted in food and water. In Europe and the USA bacteria such as salmonella, associated with industrialized farming, are prevalent. In the tropics, water-borne bacteria and cholera predominate; in coastal areas natural toxins associated with reef fish are important. Several new pathogens have recently emerged and others have become resistant to antibiotics.

Although healthy adults in industrialized countries rarely die as a result of contaminated food, the elderly and the very young are more vulnerable. Up to 15 percent of people who survive a serious gastrointestinal illness are left with a chronic condition, including kidney damage and rheumatoid arthritis. In developing countries, food-borne infections combined with chronic under-nutrition result in a large number of deaths.

Establishing the true incidence of illness caused by contaminated food and water is difficult because reporting systems are either unreliable or, in many non-industrialized countries, nonexistent. Even when studies are carried out, figures vary widely depending on what is actually measured. Access to safe water is one, imperfect, way of assessing the relative risk.

Foodborne infections

Number of cases of foodborne infections per 100,000 people
2001 or latest available

- 550 and over
- 350 – 549
- 150 – 349
- 50 – 149
- under 50

average number of bouts of diarrhea suffered each year by child under 5 years old

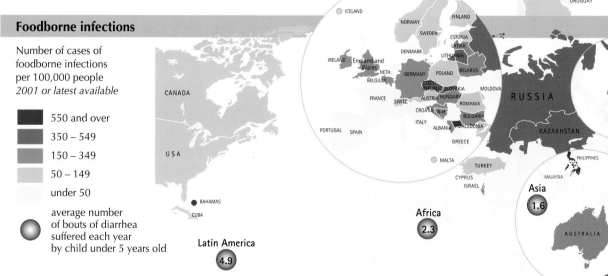

Latin America
4.9

Africa
2.3

Asia
1.6

What is clear is that illness caused by consuming contaminated food and water is a global problem. Measures needed to combat it include improved public health programs, more effective water management to increase access to safe water, and improved agricultural practices.

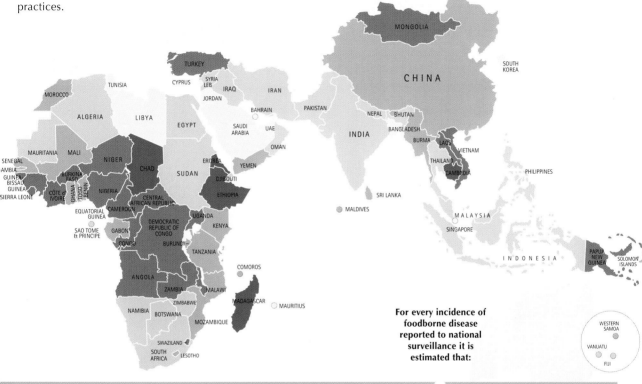

For every incidence of
foodborne disease
reported to national
surveillance it is
estimated that:

The tip of the iceberg

It is difficult to obtain accurate information on the incidence of foodborne disease, even in industrialized countries with well-established health systems. The rates shown here are as much a reflection of the efficiency of the reporting systems, as of the incidence of disease.

Even in the UK, which shows a relatively high incidence, this figure is considered to be but the tip of the iceberg.

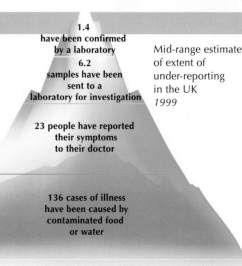

1.4
have been confirmed
by a laboratory

6.2
samples have been
sent to a
laboratory for investigation

23 people have reported
their symptoms
to their doctor

136 cases of illness
have been caused by
contaminated food
or water

Mid-range estimate
of extent of
under-reporting
in the UK
1999

Each year
Vitamin A deficiency
causes blindness in up to

500,000
children

UNDER-NUTRITION is a major public health problem. It comes in many different forms, and can be caused by an inadequate amount of food, but also by a deficiency of certain nutrients in the diet. Different types of under-nutrition often occur in the same region, and they are almost always associated with poverty.

The overall incidence of all kinds of under-nutrition is much higher in developing than in industrialized countries, with the number of underweight children highest in South Asia and Africa (see pages 12–13). This type of under-nutrition, caused by a shortage of food, is called protein-energy malnutrition, and is often associated with infectious diseases, such as measles and diarrhea. This combination is a major cause of premature death.

Other kinds of under-nutrition, such as iodine deficiency, are caused by a shortage of specific vitamins and minerals in the diet. Again, other factors often contribute to causing these deficiencies, such as infections and intestinal parasites that, for instance, reduce our ability to absorb nutrients from food.

Vitamin A deficiency (VAD) is a major public health problem, affecting 140 to 250 million pre-school children. It is a leading cause of blindness in developing countries and leaves sufferers at increased risk of infections. Iron deficiency anemia is the most common kind of micro-nutrient deficiency worldwide, and is also prevalent in industrialized countries.

Some vitamin deficiency syndromes, including rickets (vitamin D deficiency), scurvy (vitamin C deficiency), pellagra (niacin deficiency) and beri-beri (thiamine deficiency), have been largely eradicated through extensive public health programs, although they sometimes occur when people are dependent on a restricted supply of foodstuffs, such as in refugee camps. International public health bodies continue in their efforts to eradicate the remaining micronutrient deficiencies by, for instance, mass supplementation of vitamin A and iodine. However, under-nutrition caused by shortage of food continues to be a huge problem.

13%
of all people suffer from iodine deficiency

CANADA

USA

MEXICO

BAHAMAS

DOMINICAN REPUBLIC

HAITI

BELIZE
HONDURAS
GUATEMALA
EL SALVADOR   NICARAGUA

COSTA RICA
PANAMA

GUYANA

COLOMBIA

ECUADOR

PERU

BRAZIL

BOLIVIA

CHILE

Iodine Deficiency Disorders (IDD):
• affect over 740 million people
• are most prevalent in mountainous regions and river plains, where iodine has been leached from the soil by glaciers or floods
• causes goitre (enlarged thyroid gland), mental retardation in children and reduced mental capacity in adults
• are preventable by, for instance, fortifying salt with iodine

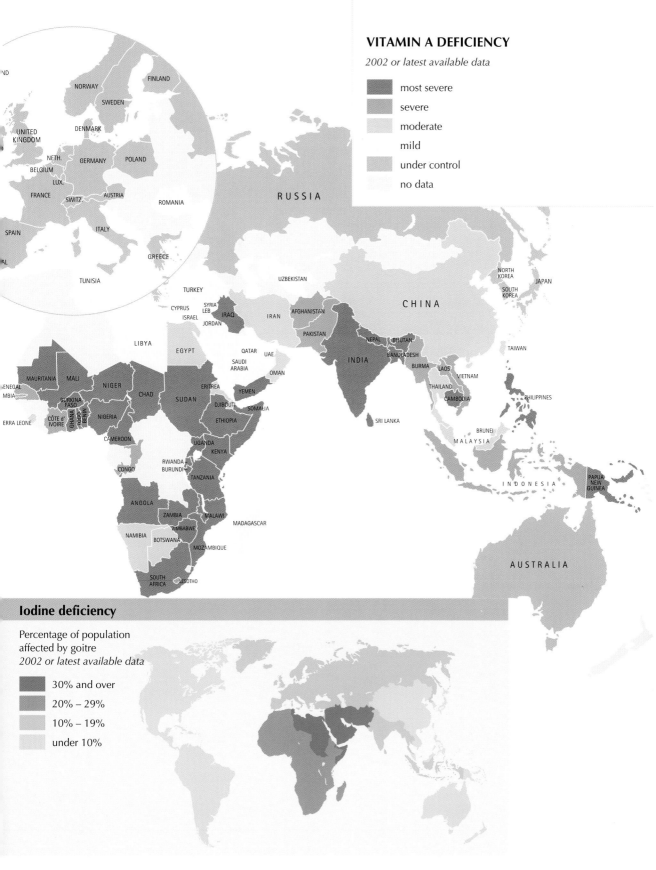

VITAMIN A DEFICIENCY

2002 or latest available data

- most severe
- severe
- moderate
- mild
- under control
- no data

Iodine deficiency

Percentage of population
affected by goitre
2002 or latest available data

- 30% and over
- 20% – 29%
- 10% – 19%
- under 10%

OVER-NUTRITION

Obesity

Percentage of people considered obese
latest available data compared with early 1980s

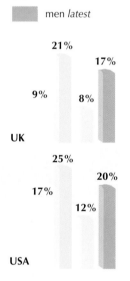

women *1980s*

women *latest*

men *1980s*

men *latest*

UK
- 21%
- 17%
- 9%
- 8%

USA
- 25%
- 20%
- 17%
- 12%

PEOPLE IN INDUSTRIALIZED countries tend to adopt sedentary lifestyles and eat more than they need. But the incidence of obesity is increasing worldwide. When countries industrialize, eating habits change and people tend to supplement their traditional diets, high in fruits, vegetables and cereals, with meat and dairy products.

Obesity can lead to diabetes, and this is rapidly becoming a worldwide epidemic. Diabetes is most common in industrialized countries but it is increasing most rapidly in areas such as the Middle East and North Africa.

Both diabetes and obesity increase the risk of coronary heart disease (CHD). Until the 1980s CHD was common in industrialized countries, but changes in lifestyle – particularly diet – and improvements in treatment has led to rates in North America, Western Europe and Australia falling. In Japan, and in countries where people have maintained their traditional, plant-based diets, rates of CHD are low, while in Russia and Eastern Europe, rates are continuing to rise. In general, premature deaths from CHD are about twice as common in men as in women, but in some regions this difference is narrowing. Indeed, diets worldwide are tending to converge – witness the growth of fast-food outlets – and in many developing countries the incidence of heart disease is increasing as a consequence.

Coronary heart disease

Number of men aged 35–74 dying each year from CHD per 100,000 population
1970–1995

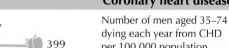

UK
- 523
- 521
- 452
- 407
- 401
- 368
- 325
- 399

Mauritius
- 272
- 263

France
Romania
- 149
- 215
- 154
- 106
- 92
- 131

| 1970 | 1980 | 1990 | 1995 |

Number of men aged 35–74 dying each year from CHD per 100,000 population
latest available data

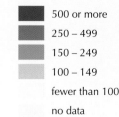

- 500 or more
- 250 – 499
- 150 – 249
- 100 – 149
- fewer than 100
- no data

death rate for women is over 50% of that for men

24

DIABETES

Percentage of people
suffering from diabetes
2000

- 5% and over
- 3.0% – 4.9%
- 1.0% – 2.9%
- under 1.0%
- no data

predicted increase
of over 90%
by 2010

ICELAND
NORWAY
FINLAND
SWEDEN
ESTONIA
LATVIA
DENMARK
LITHUANIA
UNITED
KINGDOM
NETH.
BELARUS
GERMANY
POLAND
BELGIUM
LUX.
CZECH
REPUBLIC
SLOVAKIA
UKRAINE
FRANCE
SWITZ.
AUSTRIA
SLOVENIA
HUNGARY
MOLDOVA
CROATIA
ROMANIA
B-H
SPAIN
ITALY
YUGOSLAVIA
BULGARIA
ALBANIA
MACEDONIA
GREECE

RUSSIA

KAZAKHSTAN

MONGOLIA

NORTH
KOREA
SOUTH
KOREA
JAPAN

TUNISIA
MALTA
OCCO
ALGERIA
GEORGIA
AZERBAIJAN
UZBEKISTAN
KIRGISTAN
ARMENIA
TURKMENISTAN
TAJIKISTAN
TURKEY
CYPRUS
LEBANON
SYRIA
IRAQ
IRAN
AFGHANISTAN
ISRAEL
CHINA
MOROCCO
GAZA STRIP
JORDAN
KUWAIT
BAHRAIN
PAKISTAN
ALGERIA
LIBYA
EGYPT
SAUDI
ARABIA
QATAR
UAE
NEPAL
BHUTAN
Macau I.
Hong
Kong
WESTERN
SAHARA
OMAN
INDIA
BANGLADESH
BURMA
LAOS
VIETNAM
MAURITANIA
MALI
NIGER
CHAD
ERITREA
YEMEN
SUDAN
DJIBOUTI
THAILAND
CAMBODIA
PHILIPPINES
GUAM
SENEGAL
MBIA
BURKINA
FASO
GUINEA-
BISSAU
GUINEA
CÔTE d'
IVOIRE
GHANA
TOGO
BENIN
NIGERIA
CENTRAL
AFRICAN REPUBLIC
ETHIOPIA
SOMALIA
ERRA LEONE
LIBERIA
CAMEROON
SRI LANKA
MALDIVES
EQUATORIAL
GUINEA
UGANDA
KENYA
BRUNEI
GABON
DEMOCRATIC
REPUBLIC OF
CONGO
RWANDA
BURUNDI
MALAYSIA
SINGAPORE
CONGO
TANZANIA
INDONESIA
PAPUA
NEW
GUINEA
COMOROS
EAST
TIMOR
SOLOMON
ISLANDS
ANGOLA
ZAMBIA
MALAWI
MADAGASCAR
MAURITIUS
NAMIBIA
ZIMBABWE
REUNION
BOTSWANA
MOZAMBIQUE
AUSTRALIA
SAMOA
SWAZILAND
VANUATU
FRENCH
POLYNESIA
SOUTH
AFRICA
LESOTHO
FIJI
NEW
CALEDONIA
NEW ZEALAND
25

8 FOOD AID

3.5 million tons

of food aid was distributed to 80 countries by the WFP in 2000

ENOUGH FOOD is produced in the world to feed everyone, but the availability of food varies enormously. While some countries produce a surplus, others do not produce enough to feed their populations adequately, resulting in a chronic food shortage. Acute food shortage occurs for a number of other reasons, including natural disasters such as earthquakes, environmental problems such as drought, and the impact of conflicts.

Food aid comes in many forms. Most is shipped from one country to another, but some is purchased locally, paid for by a donor. Food aid may be facilitated through international treaty arrangements such as the Food Aid Convention, and by international bodies such as the World Trade Organization, Food and Agriculture Organization, and the World Food Program (WFP). It may also be directed through non-governmental organizations such as Oxfam, or given directly from one country to another (bilateral aid).

The WFP, a UN agency, is the world's principal multilateral channel for food aid. In 2000, it distributed around 3.5 million tons of food aid to 80 countries, although not all countries considered by the FAO to be in need of relief received WFP aid. WFP aid is divided between emergency relief, longer-term relief and recovery, and development projects. However, since the 1980s its activities have become dominated by relief operations.

In 2000 the WFP spent around US$1.7 billion, donated in the form of money and food from governments, businesses and individuals, with 93 percent of its funds coming from just 12 donors. The USA distributes an additional 5 million tonnes of bilateral food aid via a number of programs aimed at developing trade links, providing emergency assistance and promoting food security and economic development. But, although the UN encourages countries to donate 0.7 percent of their GNP, of the top donors only Denmark, Netherlands, Sweden and Norway match that, with the USA contribution amounting to a mere 0.1 percent of GNP.

World Food Program

Type of food aid given
2000

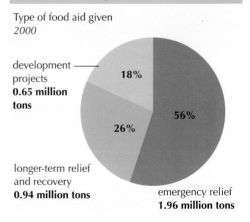

development projects
0.65 million tons — 18%

longer-term relief and recovery
0.94 million tons — 26%

emergency relief
1.96 million tons — 56%

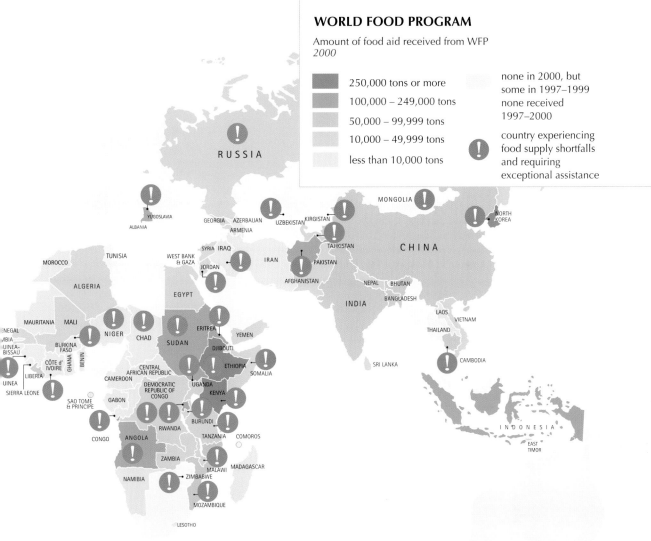

WORLD FOOD PROGRAM

Amount of food aid received from WFP
2000

■	250,000 tons or more
■	100,000 – 249,000 tons
■	50,000 – 99,999 tons
■	10,000 – 49,999 tons
□	less than 10,000 tons

none in 2000, but
some in 1997–1999

none received
1997–2000

! country experiencing
food supply shortfalls
and requiring
exceptional assistance

RUSSIA

MONGOLIA

YUGOSLAVIA
ALBANIA
GEORGIA AZERBAIJAN
ARMENIA
UZBEKISTAN KIRGISTAN
NORTH
KOREA
SYRIA IRAQ
WEST BANK
& GAZA
JORDAN
IRAN
TAJIKISTAN
CHINA
PAKISTAN
AFGHANISTAN
MOROCCO
TUNISIA
NEPAL BHUTAN
ALGERIA
EGYPT
INDIA
BANGLADESH
MAURITANIA MALI
NIGER CHAD
ERITREA
YEMEN
LAOS
VIETNAM
NEGAL
MBIA
UINEA-
BISSAU
BURKINA
FASO
SUDAN
DJIBOUTI
THAILAND
SRI LANKA
CÔTE d'
IVOIRE
GHANA
BENIN
CENTRAL
AFRICAN REPUBLIC
ETHIOPIA
SOMALIA
CAMBODIA
LIBERIA
CAMEROON
UGANDA
UINEA
SIERRA LEONE
SAO TOME
& PRINCIPE
GABON
DEMOCRATIC
REPUBLIC OF
CONGO
KENYA
CONGO
RWANDA
BURUNDI
TANZANIA
COMOROS
INDONESIA
ANGOLA
ZAMBIA
MALAWI
MADAGASCAR
EAST
TIMOR
NAMIBIA
ZIMBABWE
MOZAMBIQUE
LESOTHO

Top twelve donors

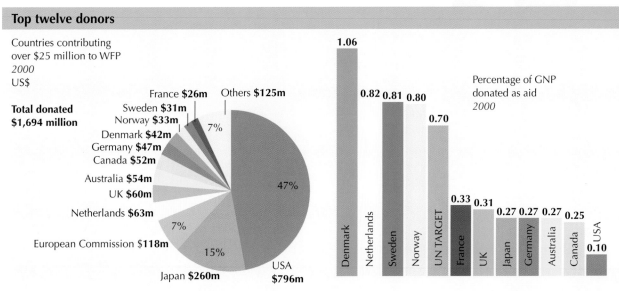

Countries contributing
over $25 million to WFP
2000
US$

**Total donated
$1,694 million**

France **$26m**
Sweden **$31m**
Norway **$33m**
Denmark **$42m**
Germany **$47m**
Canada **$52m**
Australia **$54m**
UK **$60m**
Netherlands **$63m**
European Commission **$118m**
Japan **$260m**

Others **$125m**
7%

47%

15%

7%

USA
$796m

Percentage of GNP
donated as aid
2000

Denmark 1.06
Netherlands 0.82
Sweden 0.81
Norway 0.80
UN TARGET 0.70
France 0.33
UK 0.31
Japan 0.27
Germany 0.27
Australia 0.27
Canada 0.25
USA 0.10

9 FOOD AID AS POWER

WELL-MANAGED FOOD AID clearly benefits people suffering from a shortage of food. Conversely, poorly managed food aid (which may be culturally or nutritionally inappropriate) contributes to long-term food insecurity by discouraging local food production, depressing prices, and disrupting local markets.

Large quantities of food aid may change patterns of world trade and result in countries becoming dependent on food imports. For example, in the late 1950s one third of wheat traded on international markets came from the USA in the form of food aid. Many developing countries in Africa, Asia and Latin America became major importers of cheap US wheat, with drastic results for their domestic agriculture. In Colombia, for example, production of wheat, potatoes and barley, collapsed.

Where a large amount of aid is given to a country at a time of crisis, there is the danger of food aid becoming "institutionalized" and continuing longer than is desirable. In the late 1990s, over 50 percent of the food imported into Ethiopia, Rwanda and Uganda still came in as aid, some years after the crises that had initially prompted the aid.

Food aid is often used by the donor country as a way of influencing the politics in the recipient country. For example, it may be contingent on the development of particular policies, such as the establishment of democratic institutions. Threatening to withdraw aid is also an effective way of exerting influence. Foreign officials may even take up posts in the policy-making structures of the recipient country.

Recipient governments in countries where local food production is expensive may use food aid to support their own economic targets. For example, extensive food aid will depress local food prices, keeping wages low and enabling the cheap production of export products that generate foreign exchange.

While it is likely that there will always be acute periods when emergency food aid is required, in an ideal world long-term food insecurity, and hence the need for protracted relief and development food aid, would be eliminated. In the meantime, mechanisms for donating food aid need to be adapted and reformed.

Greater flexibility is needed in the type of food aid given. Different foods are appropriate in different situations, and sometimes money that enables recipient countries to purchase their own food is preferable. There also needs to be greater stability in supply and sufficient resources for developing food aid that promotes long-term food security and reduces the dependence of recipient countries on imported food aid.

$32 billion

of food aid was distributed to needy people within the USA in 2001

Top ten recipients of WFP aid

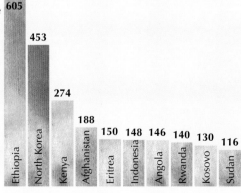

Countries receiving the most WFP aid
2000
thousand tons

A large proportion of food aid goes to a small number of recipients. In 2000 two-thirds of WFP aid went to just 10 countries.

Some of this aid was in response to crises, such as occurred in Kosovo, Yugoslavia. Some was given in response to chronic food shortage, such as in North Korea and Afghanistan.

While provision of food aid through multilateral organizations is intended to ensure apolitical allocation of resources, evidence suggests that in some cases emergency aid is directed to recipient countries specified by the donor.

US food aid

As well as donating nearly half of the WFP's budget, the USA distributes its own international food aid via a number of programs, each of which has a specific function.

Aid provided under the **Public Law 480 program, Title I**, is intended to develop trade, while that under **Title II** is for emergency assistance and for promoting food security and economic development.

Section 416(b) aid was authorized by the Agricultural Act of 1949 and allows the USA to off-load surplus agricultural products. The **Food for Progress** scheme is for long-term development work.

Over half of all US food aid is directed at Asia and countries of the former Soviet Union (see below).

The amount and type of aid provided under Section 416(b) varies according to world market prices and the amount of surplus food produced. This introduces a level of uncertainty into the supply of aid. Until recently, food aid organized by the European Commission, was also linked to agricultural surpluses, but the **1996 Food Aid Regulation** introduced more flexibility, allowing financial aid to be donated as well as food, where appropriate.

Type of aid
2000

emergency relief 30%

aid for trade 21%

agricultural surplus 44%

development aid 5%

Aid for trade

Russia 62%

Philippines 15%

Morocco 9%

Sri Lanka 4%

Uzbekistan 4%

Peru 3%

Angola 2%

Jamaica 1%

Recipients of US bilateral food aid
2000

Total donated: 4.96 million tonnes

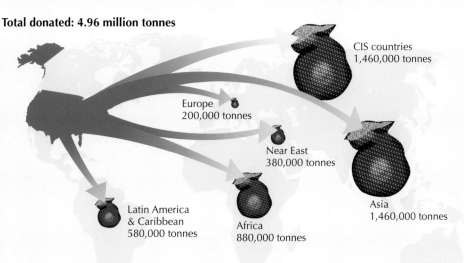

CIS countries
1,460,000 tonnes

Europe
200,000 tonnes

Near East
380,000 tonnes

Asia
1,460,000 tonnes

Africa
880,000 tonnes

Latin America & Caribbean
580,000 tonnes

Not all US food aid is free. **Public Law 480, Title I** food aid (shown above) is traded for US dollars or for local currency, often on long credit terms and at low interest rates. Priority is given to countries that take steps to improve food security and promote broad-based, equitable and sustainable development.

Crucially, these countries have to demonstrate their potential to become commercial markets for US agricultural commodities. Much of this kind of aid goes to Russia, since trade with a healthy Russian economy benefits the USA.

FARMING

"Sixty percent of the international food chain is controlled by just 10 companies, involved in seed, fertilizers, pesticides, processing and shipments."
– David C Korten, *When Corporations Rule the World,* 1995

2

303078

THE MECHANIZATION of farming began in the 19th century with the development of the threshing machine, the seed drill and the reaper. But the major breakthrough was the application of the internal combustion engine in the early 20th century. This produced the farm tractor and associated machinery as well as the combine harvester, and machines such as corn dryers, shearers and milking machines.

Since the 1960s tractors have played an increasing part in agriculture worldwide. They have also become ever more powerful: even though the total number of tractors peaked in 1990, those used in 2000 produce more horse power than those of ten years earlier.

The use of tractors tends to mirror economic developments, with the least mechanized countries in Sub-Saharan Africa, South America and Asia. In around two-thirds of these countries there are fewer than 50 tractors per 1,000 agricultural workers and their dependants. In Rwanda, the least mechanized country, there is only 1 tractor for every 40,000 people in the agricultural population.

In some countries, mechanization has occurred at a furious rate. These range from African countries such as Nigeria and Burkina Faso, to those in the Middle East, Asia, and eastern Europe. Some terrains, such as the terraced fields of Southeast Asia, are unsuitable for tractors, yet even in countries such as Indonesia and Thailand there has been a notable increase in their use. In Japan the number of tractors increased by nearly *12,000 percent* between 1965 and 1998, although these are relatively small machines, suitable for use on the wet rice fields, and a far cry from the giants of the North American prairies.

The USA, Canada and several European countries have more tractors than people involved in agriculture. However, some of the most technically advanced agriculture in the world is found in the Netherlands and in New Zealand, where the greater importance of animal production means that fewer tractors are required than in primarily arable areas.

Increased tractor power

Average tractor horse power
1920s, 1950s and 2000

2000 240hp

1950s 24hp

1920s 12hp

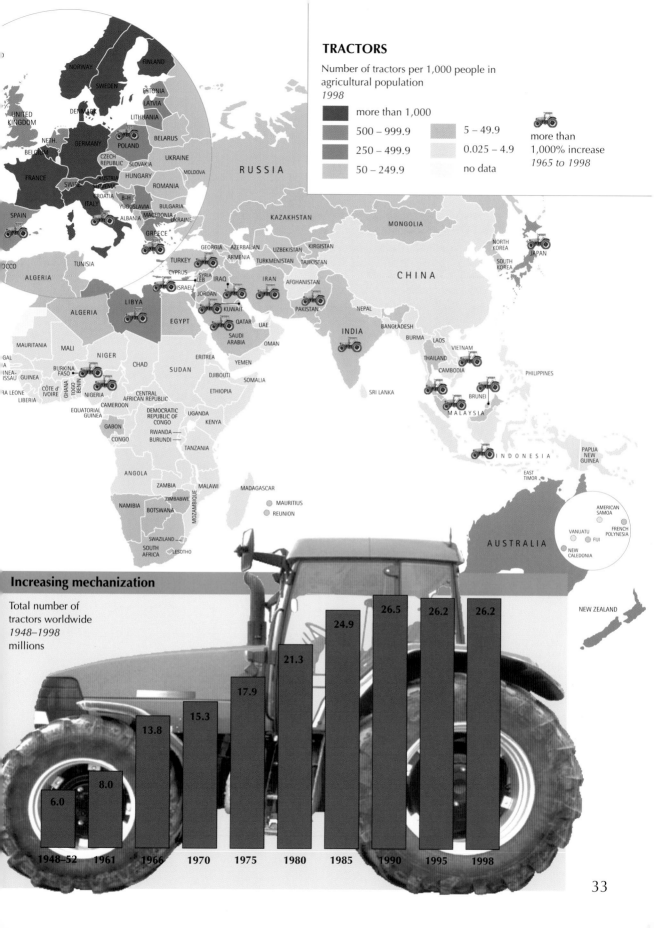

TRACTORS

Number of tractors per 1,000 people in agricultural population
1998

more than 1,000	
500 – 999.9	5 – 49.9
250 – 499.9	0.025 – 4.9
50 – 249.9	no data

more than 1,000% increase
1965 to 1998

Increasing mechanization

Total number of tractors worldwide
1948–1998
millions

1948–52	1961	1966	1970	1975	1980	1985	1990	1995	1998
6.0	8.0	13.8	15.3	17.9	21.3	24.9	26.5	26.2	26.2

33

11 ANIMAL FEED

75%

of agricultural
land in the EU is
used for growing
animal feed

MOST PEOPLE in the world eat a mainly vegetarian diet. In affluent countries, however, the consumption of meat, eggs, milk and dairy products has steadily increased since the 1950s.

Technological advances in crop production, especially in North America and Western Europe, has led to a supply of cheap and plentiful surplus grains and pulses. This, combined with a consumer demand for more beef, pork, poultry milk and dairy products, has provided major financial opportunities for livestock farmers. While meat production may make a great deal of economic sense, at least in the short run, it does not make ecological and nutritional sense in the long run.

Livestock farmers use their animals to transform the cheap and plentiful raw material of grain into the relatively scarce, expensive and profitable product of meat. Around 95 percent of the world's commercial soybean harvest, and a third of commercial fish catches, are consumed by animals rather than humans. Much of the meat produced in the industrialized countries comes from "intensive feedlots". In this system, animals have relatively little freedom to roam and graze naturally, but are provided with large amounts of feed designed to maximize growth, so that they reach their target weight as quickly as possible, or yield the maximum amount of milk or eggs.

Intensive feed-lot production of meat, eggs and milk is much more environmentally demanding than traditional forms of arable farming. Farm animals now consume increasing amounts of land, energy, and water.

930 kg

180 kg

The grain drain

Amount of grain needed to feed one person for one year on a meat-based diet or on a grain-based diet.

meat-based diet **grain-based diet**

Feeding animals to feed people is a costly way to produce food. A quarter of the earth's landmass is used as pasture for livestock farming. Half of all US farmland, directly or indirectly, is devoted to beef production. In the EU, 75 percent of agricultural land is used for growing animal feed. Livestock production contributes extensively to soil erosion and desertification, with 85 percent of topsoil loss in the USA directly attributable to livestock ranching.

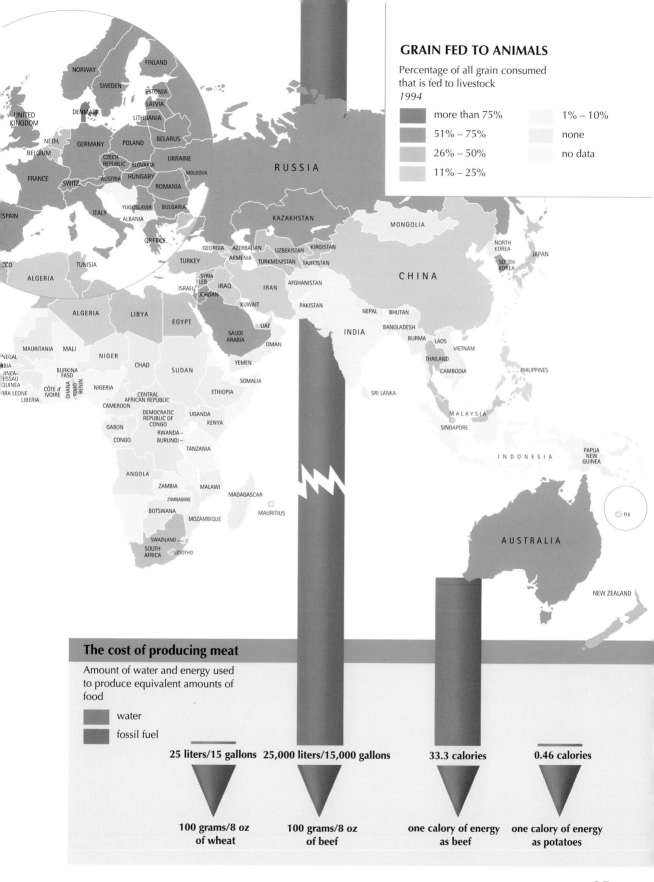

GRAIN FED TO ANIMALS

Percentage of all grain consumed
that is fed to livestock
1994

- more than 75%
- 51% – 75%
- 26% – 50%
- 11% – 25%
- 1% – 10%
- none
- no data

The cost of producing meat

Amount of water and energy used
to produce equivalent amounts of
food

- water
- fossil fuel

25 liters/15 gallons	25,000 liters/15,000 gallons	33.3 calories	0.46 calories
100 grams/8 oz of wheat	100 grams/8 oz of beef	one calorie of energy as beef	one calorie of energy as potatoes

35

BSE outside the UK

Total number of cases of BSE *up to May 2002* selected countries

France
1,109

Ireland
1,021

Portugal
657

Switzerland
416

Germany
153

Spain
84

MAD COW DISEASE is the popular name for bovine spongiform encephalopathy (BSE), a condition that can affect the brains of cattle. It first emerged in the mid-1980s in the UK, and in 1993 there were over 37,000 cases. Infected animals show loss of coordination and a change of temperament. The disease is untreatable and the animals have to be destroyed. When the disease was first identified in cattle no-one could tell whether or not it could be passed from animals to humans, but in March 1996 an entirely novel BSE-like disease was identified in humans, which was termed "variant Creuzfeld-Jakob Disease" or vCJD.

BSE is thought to be caused by a protein called a prion. Prions are thought to have entered the cattle-feed chain when animal carcasses were rendered down into meat and bone meal (MBM) and incorporated into cattle feed without being adequately decontaminated. If this is the case, the disease is clearly a consequence of large-scale industrialization of food production.

BSE has subsequently emerged in several other countries, after British MBM was exported for use in cattle feed. Despite regulations to control the spread of BSE – first introduced in 1989 in the UK, and subsequently strengthened and extended to many other countries – it has not yet been eradicated.

USA

COLOMBIA

ECUADOR

BSE in the UK

Confirmed cases of BSE
1987–2001

Total cases: 180,725

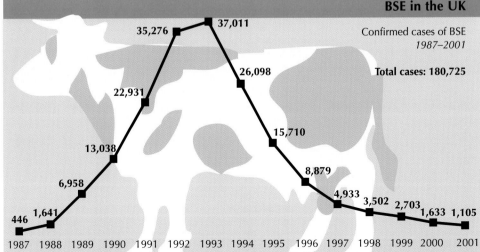

Year	Cases
1987	446
1988	1,641
1989	6,958
1990	13,038
1991	22,931
1992	35,276
1993	37,011
1994	26,098
1995	15,710
1996	8,879
1997	4,933
1998	3,502
1999	2,703
2000	1,633
2001	1,105

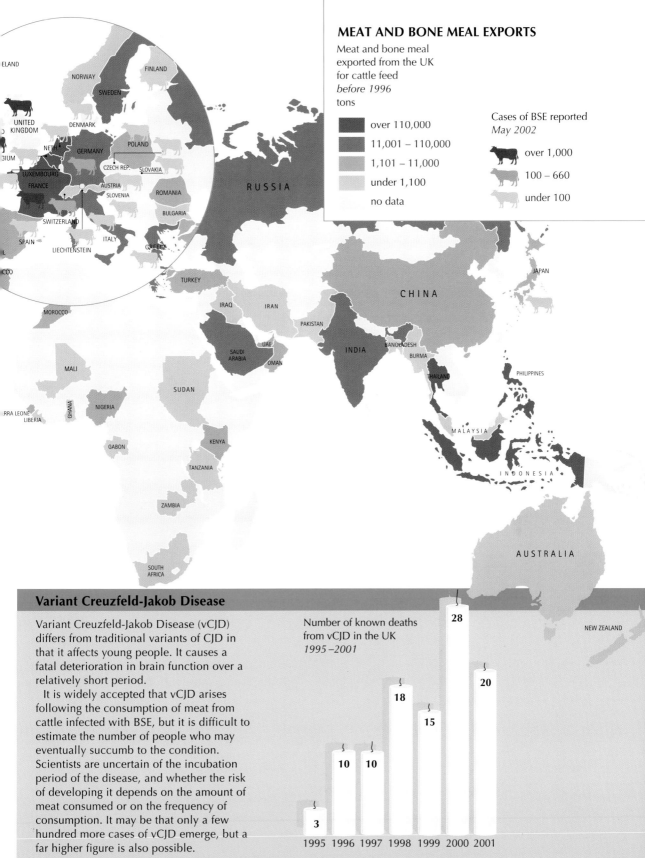

MEAT AND BONE MEAL EXPORTS

Meat and bone meal
exported from the UK
for cattle feed
before 1996
tons

- over 110,000
- 11,001 – 110,000
- 1,101 – 11,000
- under 1,100
- no data

Cases of BSE reported
May 2002

- over 1,000
- 100 – 660
- under 100

ELAND
NORWAY
FINLAND
SWEDEN
UNITED KINGDOM
DENMARK
GERMANY
POLAND
3IUM
NETH.
CZECH REP.
SLOVAKIA
LUXEMBOURG
FRANCE
AUSTRIA
SLOVENIA
ROMANIA
SWITZERLAND
LIECHTENSTEIN
ITALY
BULGARIA
SPAIN
GREECE
L
CCO
TURKEY
MOROCCO
IRAQ
IRAN
PAKISTAN
MALI
SAUDI ARABIA
UAE
OMAN
SUDAN
RRA LEONE
LIBERIA
GHANA
NIGERIA
GABON
KENYA
TANZANIA
ZAMBIA
SOUTH AFRICA

RUSSIA
CHINA
JAPAN
INDIA
BANGLADESH
BURMA
THAILAND
PHILIPPINES
MALAYSIA
INDONESIA
AUSTRALIA
NEW ZEALAND

Variant Creuzfeld-Jakob Disease

Variant Creuzfeld-Jakob Disease (vCJD) differs from traditional variants of CJD in that it affects young people. It causes a fatal deterioration in brain function over a relatively short period.

It is widely accepted that vCJD arises following the consumption of meat from cattle infected with BSE, but it is difficult to estimate the number of people who may eventually succumb to the condition. Scientists are uncertain of the incubation period of the disease, and whether the risk of developing it depends on the amount of meat consumed or on the frequency of consumption. It may be that only a few hundred more cases of vCJD emerge, but a far higher figure is also possible.

Number of known deaths
from vCJD in the UK
1995 –2001

Year	Deaths
1995	3
1996	10
1997	10
1998	18
1999	15
2000	28
2001	20

INDUSTRIAL FARMING

Battery chickens are kept up to nine birds to a cage in the USA, and up to five per cage in the European Union.

INDUSTRIAL FARMING is a system of rearing animals using intensive "production-line" methods that aim to maximize the amount of meat produced, while minimizing costs. Encouraged by governments as a response to food shortages and the need to raise productivity after World War II, industrial farming has dominated animal agriculture in the European Union (EU) and North America since the 1950s.

Intensive farming inhibits animals from behaving naturally, often causing them pain and serious health problems. In some countries, campaigns against industrial farming methods have influenced public opinion and government agencies. The EU, for example, has adopted measures to ban battery cages for egg-laying hens by 2012 and sow gestation crates (stalls) by 2013. However, in the USA, no real improvements are in sight.

Intensive farming is also harmful to human health. Animals in cramped conditions easily catch and transmit bacteria, which may then be passed to humans. Farmers routinely use antibiotics to combat infectious diseases, but in so doing may be contributing to growing antibiotic resistance among humans.

The conversion of cereals and legumes into meat is a nutritionally inefficient and costly

Pig farming

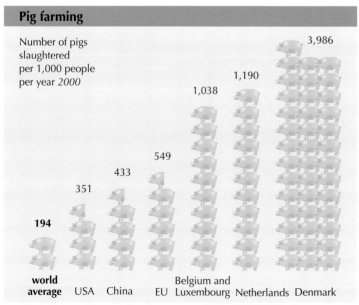

Number of pigs slaughtered per 1,000 people per year *2000*

194	351	433	549	1,038	1,190	3,986
world average	USA	China	EU	Belgium and Luxembourg	Netherlands	Denmark

way of feeding rapidly increasing populations. Yet countries of Europe and North America are not alone in rearing large numbers of animals for meat. Meat production is increasing in South America, Southeast Asia and elsewhere. To do this farmers are adopting intensive-farming methods, often encouraged by foreign investors seeking new markets for intensive livestock-rearing equipment, and their own governments seeking outlets for crop surpluses. Rural livelihoods are undermined as small-scale farmers are unable to compete with large intensive pig or poultry units. The

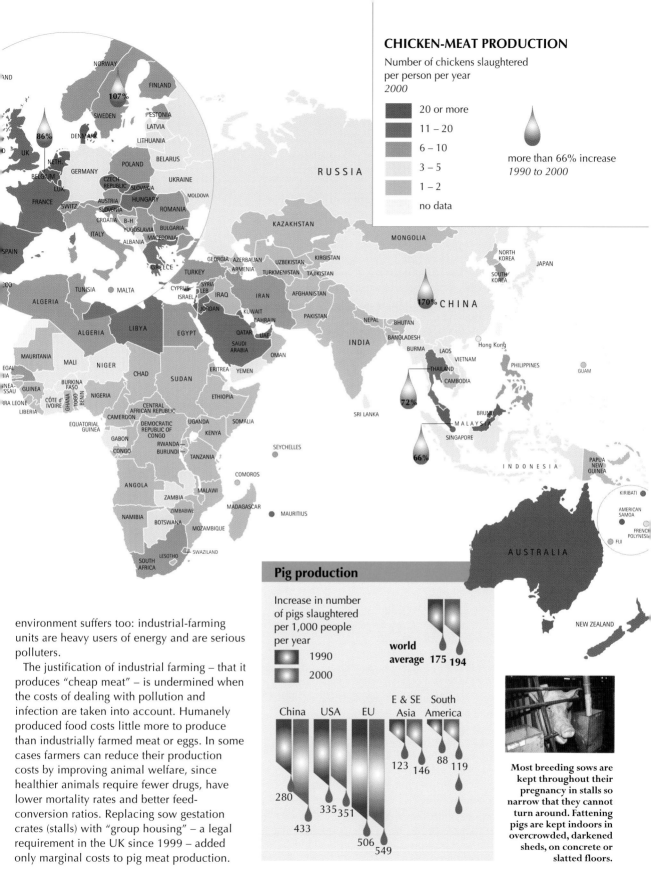

CHICKEN-MEAT PRODUCTION

Number of chickens slaughtered per person per year
2000

- 20 or more
- 11 – 20
- 6 – 10
- 3 – 5
- 1 – 2
- no data

more than 66% increase
1990 to 2000

Pig production

Increase in number of pigs slaughtered per 1,000 people per year

- 1990
- 2000

world average 175 194

	China	USA	EU	E & SE Asia	South America
1990	280	335	351	123	88
2000	433		506 / 549	146	119

Most breeding sows are kept throughout their pregnancy in stalls so narrow that they cannot turn around. Fattening pigs are kept indoors in overcrowded, darkened sheds, on concrete or slatted floors.

environment suffers too: industrial-farming units are heavy users of energy and are serious polluters.

The justification of industrial farming – that it produces "cheap meat" – is undermined when the costs of dealing with pollution and infection are taken into account. Humanely produced food costs little more to produce than industrially farmed meat or eggs. In some cases farmers can reduce their production costs by improving animal welfare, since healthier animals require fewer drugs, have lower mortality rates and better feed-conversion ratios. Replacing sow gestation crates (stalls) with "group housing" – a legal requirement in the UK since 1999 – added only marginal costs to pig meat production.

Wheat harvest at the Great Plains Research Station in Colorado, USA.

UNTIL THE EARLY 1980s agricultural research and development (R&D) in industrialized countries was largely publicly funded and freely available to farmers. Research was carried out at all levels – from basic science, through to its application on the farm and ways of improving productivity. In recent years, however, local food surpluses have meant that priorities have switched to improving the performance of the products after they have left the farm – storing them safely, and enhancing their shelf-life and suitability for processing.

The application of science to agriculture is now seen as the domain of the market-led private sector, with research focusing largely on the needs of capital-intensive farming. The focus of this research is also changing. In the USA, for example, 80 percent of private-sector funding in 1960 was spent on agricultural machinery and on food-processing, whereas funding is now aimed at research into plant breeding and veterinary products.

Almost all research in developing countries is carried out by public bodies. The financial rewards are insufficient for private companies, although the development of new strains of rice may attract future private funding. Between 1976 and 1991 research expanded most rapidly in China, and in countries in Asia, the Pacific Rim and North Africa, but much more slowly in Sub-Saharan Africa, Latin America and the Caribbean. More recently, economic factors may have led to reductions in these growth rates.

Research into plant breeding and farming technologies tends to be specific to a region or climate. Although some technologies suitable for tropical conditions are being developed in temperate countries, with only 28 percent of all publicly funded R&D taking place in the tropics, the needs of some of the least productive and poorest countries are being neglected. Although R&D in developing countries has traditionally been funded by international bodies, this ear-marked funding has declined since the mid-1980s, with a greater emphasis being placed on supporting a country's economic infrastructure, rather than on specific projects.

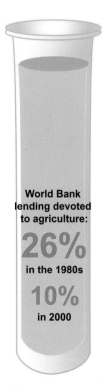

World Bank lending devoted to agriculture:

26%
in the 1980s

10%
in 2000

USA

MEXICO

GUATEMALA   HONDURAS

COSTA RICA
PANAMA

COLOMBIA

ANTIGUA
& BARBUDA

ST KITTS & NEVIS

$1,087m $1,583m $1,947m

Latin America and Caribbean

BRAZIL

CHILE

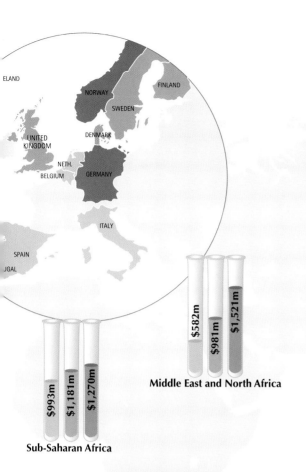

ELAND

NORWAY
FINLAND

SWEDEN

UNITED KINGDOM
DENMARK

NETH.
BELGIUM
GERMANY

ITALY

SPAIN

JGAL

EXTENT OF PUBLIC FUNDING

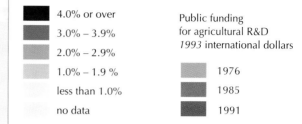

Public expenditure on R&D
as a percentage of agriculturalGDP
1995
selected countries

■	4.0% or over
■	3.0% – 3.9%
■	2.0% – 2.9%
■	1.0% – 1.9 %
	less than 1.0%
	no data

Public funding
for agricultural R&D
1993 international dollars

- 1976
- 1985
- 1991

JAPAN

CHINA

$582m **$981m** **$1,521m**

Middle East and North Africa

$993m **$1,181m** **$1,270m**

Sub-Saharan Africa

$709m **$1,396m** **$2,063m**

China

$1,321m **$2,453m** **$4,619m**

Asia and Pacific

AUSTRALIA

NEW ZEALAND

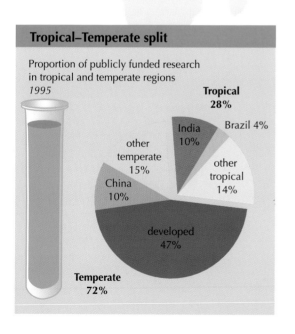

Tropical–Temperate split

Proportion of publicly funded research
in tropical and temperate regions
1995

**Tropical
28%**

India
10%

Brazil 4%

other
temperate
15%

other
tropical
14%

China
10%

developed
47%

**Temperate
72%**

Public–Private split

Proportion of funding
from public and private
organizations
1995
international dollars (1993)

private
6%

private
51.5%

public
48.5%

public
94%

**Developed countries
total: $21 billion**

**Developing countries
total: $12.2 billion**

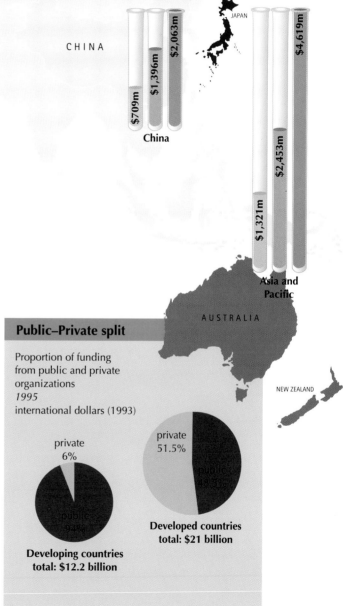

15 | GENETIC MODIFICATION

Arguments for and against

Advantages:
- GM crops will increase yields and help feed expanding populations.
- Crops can be modified to grow in dry areas or where salinity is high.
- Levels of nutrients such as betacarotene can be increased in staple crops.
- Vaccines can be cheaply produced in plants.

Disadvantages:
- Research is not primarily aimed at the needs of poor.
- Farmers will become dependent on biotech companies for supplies of expensive GM seeds.
- GM genes might "escape" and GM crops grow where they are not wanted.
- GM crops grown in the developed world might be substituted for crops traditionally imported from developing countries.

THE FIRST GENETICALLY MODIFIED (GM) plant was produced in 1984. Since then over 60 different plant species, including all the major food crops and trees, have been genetically modified and tested in field trials. Animals – including cattle, sheep, fish and insects – have also been genetically modified.

Genetic modification differs from traditional plant breeding because it allows genes from one species to be moved into another. Conventional breeding only involves the same species or very closely related species, which restricts the gene pool and available characteristics that can be transferred into the new plant.

Most genetic modifications to plants involve taking genes from bacteria, viruses and other plants. Sometimes genes have been transferred from animals. Fish genes that provide tolerance to cold have been moved into strawberries, for example, but these are unlikely to be commercialized because of consumer resistance.

The use of human genes in plants to produce proteins for use as drugs is likely to be more acceptable.

The transferred genes play different roles. Some change the character of the plant, by making it insect-resistant, for example. Some genes are used to "switch on" another introduced gene. Other genes are introduced to act on a natural gene by, for example, switching the natural gene off and thereby delaying fruit from ripening.

Because the genetic modification technique is imprecise and only works in a small number of cases, a "marker" gene is usually included. This codes for a characteristic such as antibiotic resistance, which can then be identified in the laboratory and used to indicate those cells that have been successfuly modified.

Biotechnology companies are investing heavily in research into GM food, and they want to ensure that they receive a financial return on their investment by controlling who has access to genes and GM plants. They are doing this in two ways.

First, they are claiming patent protection for the genes they discover and the GM crops and seed they produce. This gives them a monopoly on the commercial exploitation of their "invention" for 20 years and allows them to charge royalties or license fees for its use. Farmers growing plants from patented seed will have to pay royalties on any seed they buy or keep for re-sowing, pushing up their costs and excluding the poorest farmers from using GM seed. Many people have raised objections to this on the grounds that discoveries about nature cannot not be claimed as inventions.

Secondly, the biotechnology companies are exploring Genetic Use Restriction Technology (GURTs), which ensures that farmers using their seed are forced to purchase additional chemicals that need to be applied before the seed or plant will function.

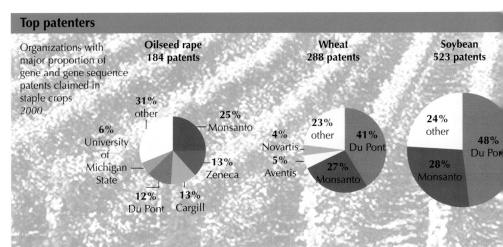

Top patenters

Organizations with major proportion of gene and gene sequence patents claimed in staple crops 2000

Oilseed rape 184 patents
- 31% other
- 25% Monsanto
- 6% University of Michigan State
- 13% Zeneca
- 12% Du Pont
- 13% Cargill

Wheat 288 patents
- 23% other
- 41% Du Pont
- 4% Novartis
- 5% Aventis
- 27% Monsanto

Soybean 523 patents
- 24% other
- 48% Du Pont
- 28% Monsanto

Keeping control

Genetic Use Restriction Technologies (GURTs) allow companies to control who uses their products. Plants can be modified so that certain characteristics are not activated until a specific chemical is applied.

Terminator technology makes a plant's seeds sterile so they can only be re-sown by purchasing specific chemicals.

Traitor technology controls characteristics such as time of flowering or disease resistance so that they can only be switched on by the use of chemicals. Many biotechnology companies promised a pause in the development of terminator technology following the 1999–2000 international campaign against it which drew attention to its impact on farmers in developing countries.

However, because terminator seed is sterile, it is promoted as a way of preventing genes in GM crops spreading in the environment.

Super salmon

Experiments have been conducted on salmon in which a gene is modified so that the fish grow more quickly.

While this has obvious commercial benefits, critics question the potential risks for humans and for the environment if such salmon were to escape into the wild.

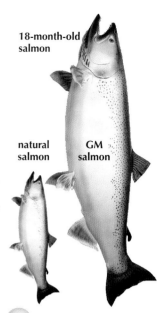

18-month-old salmon

natural salmon

GM salmon

Golden rice

"Golden rice" was announced in January 2000 as an example of how GM crops can help the developing world.

Polished rice makes up the majority of many people's diet in Asia and Southeast Asia but is low in beta-carotene – which is absorbed by the body and converted into Vitamin A. As a result, Vitamin A deficiency (VAD) is common in many developing countries, affecting between 0.25 and 0.5 million pre-school children a year. It causes blindness and, in many cases, death.

Golden rice has been genetically modified to increase its beta-carotene content. The companies involved have given up their patents for use in poor countries.

Critics question how effectively the beta-carotene can be absorbed, and point out that the yellowy-brown rice may not be readily accepted by people who judge the quality of rice by its whiteness. An education program will be needed to encourage people to eat golden rice, and it has been suggested that encouraging the growing and consumption of more fruit and vegetables, eggs and cheese would have wider nutritional benefits. Improving incomes, education and sanitation, or giving supplements to those in need could prevent deaths and disability arising from reasons other than vitamin A deficiency.

Health fears

- New combinations of genes may have unintended effects on the human body, such as producing unexpected toxins.

- New allergenic proteins might be created.

- The use of antibiotic marker genes might increase antibiotic resistance in disease-causing organisms.

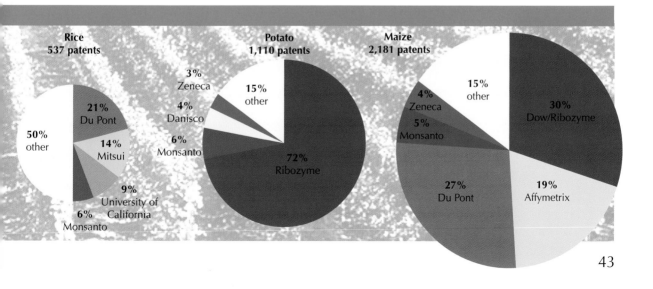

Rice
537 patents

50% other
21% Du Pont
14% Mitsui
9% University of California
6% Monsanto

Potato
1,110 patents

3% Zeneca
15% other
4% Danisco
6% Monsanto
72% Ribozyme

Maize
2,181 patents

15% other
4% Zeneca
5% Monsanto
30% Dow/Ribozyme
27% Du Pont
19% Affymetrix

GENETICALLY MODIFIED CROPS were first grown commercially in the USA in 1996. Soybean, maize, cotton and canola/oilseed rape that are tolerant to to the broad-spectrum herbicides glyphosate and glufosinate are now grown commercially, as is maize and cotton that is also resistant to some caterpillars.

Since 1996 cultivation of GM crops has increased dramatically in the USA, but more slowly elsewhere. This is because the GM crops initially developed – herbicide-tolerant soybean and insect-resistant maize and cotton – were designed for the US farming system, which grows commodity crops on a vast scale. In Argentina, Australia and Canada, where there are similar agricultural practices, genetic modifications that make crop-management easier are attractive, and GM crops are grown commercially on a limited scale. Elsewhere in the world, the benefits of GM crops are not as immediately obvious.

Consumer resistance has also affected the wider adoption of GM crops. In Europe, public fears over safety has been widely voiced and has effectively halted the commercial production of GM crops from spreading around the world. Food producers have been forced by public opinion to source non-GM ingredients, and countries such as Brazil, which have become important suppliers of non-GM soybeans to the European market, are holding back from introducing GM crops.

In addition to the concerns about how safe it is to eat foods containing GM products, there are fears about the effect on the environment. Herbicide-tolerant crops that allow for more efficient weed removal, may adversely affect farmland species that depend on weeds for food. The use of insect-resistant crops may reduce the use of insecticides in some circumstances, but the toxin in the GM crops may harm non-target species and have a damaging effect on the food chain.

Herbicide tolerance and insect resistance remains the focus of research into new GM crops, with these traits being introduced in a wider range of crops. Disease-resistant crops, and oilseed crops with altered fatty-acid profiles that are better suited to processing, are likely to form the next generation of GM crops. Researchers are also seeking to produce food products with altered starch and sugar content, and reduced levels of allergens.

GM crops

Type of crop as percentage of total GM crops *2000*

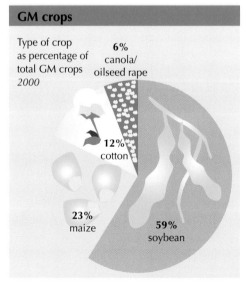

6% canola/oilseed rape

12% cotton

23% maize

59% soybean

GM traits

Type of trait as percentage of total GM crops *2000*

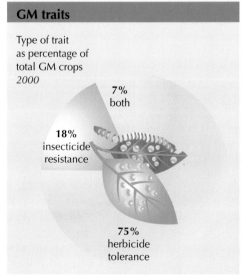

7% both

18% insecticide resistance

75% herbicide tolerance

GM crops worldwide

1997–2001
hectares

11 million hectares
1997

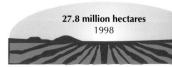

27.8 million hectares
1998

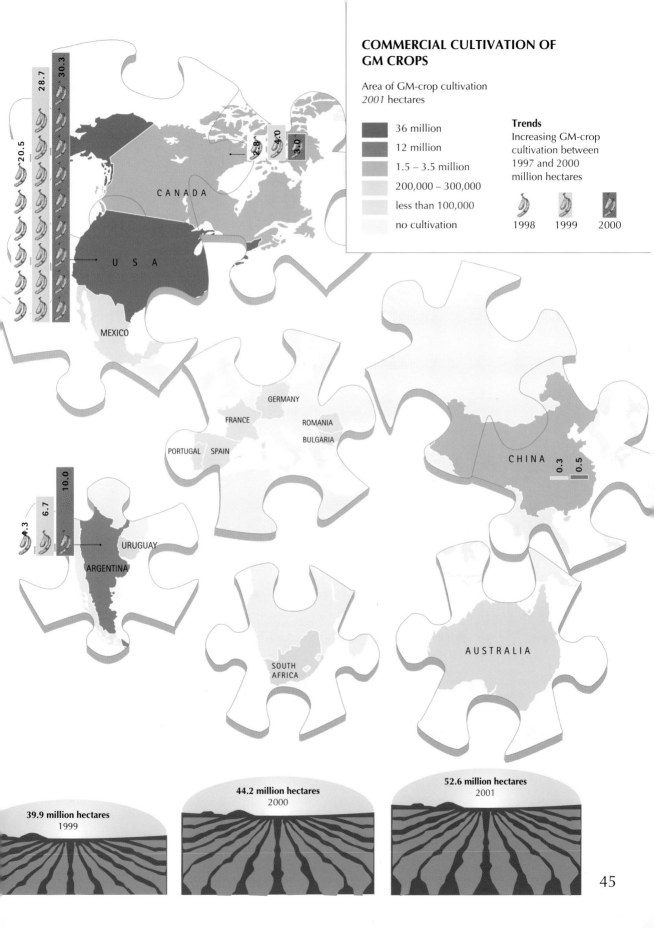

COMMERCIAL CULTIVATION OF GM CROPS

Area of GM-crop cultivation
2001 hectares

- 36 million
- 12 million
- 1.5 – 3.5 million
- 200,000 – 300,000
- less than 100,000
- no cultivation

Trends
Increasing GM-crop cultivation between 1997 and 2000 million hectares

1998 1999 2000

20.5
28.7
30.3

2.8 4.0 3.0

CANADA

U S A

MEXICO

GERMANY
FRANCE
ROMANIA
BULGARIA
PORTUGAL SPAIN

CHINA 0.3 0.5

4.3
6.7
10.0

URUGUAY
ARGENTINA

SOUTH
AFRICA

AUSTRALIA

39.9 million hectares
1999

44.2 million hectares
2000

52.6 million hectares
2001

45

Types of pesticides sold

2001
US$

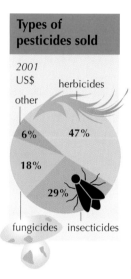

- herbicides **47%**
- other **6%**
- **18%**
- **29%**
- fungicides
- insecticides

HERBICIDES, FUNGICIDES and insecticides – collectively known as pesticides – are manufactured substances used to control weeds, fungi and insects that may reduce crop yields and damage crop quality. Since the 1950s, sales have dramatically increased and they are predicted to continue to do so, as agricultural practices in developing countries are modernized.

Pesticides are aggressively promoted worldwide, in particular in Asia and Latin America. But although they appear to provide a short-term increase in productivity, estimates of their value to agriculture rarely take into account their true cost. This includes damage to the environment and to human health, the development of pesticide-resistant pests, and the expense of testing for residues and disposing of unwanted chemicals.

Pesticides can have a devastating impact on human health, and kill an estimated 20,000 agricultural workers every year. In Benin, for example, where cotton is treated with the insecticide endosulfan, around 70 people died in 1999 because food crops were contaminated, proper protective clothing was unavailable, and farmers were not adequately informed about the products they used.

Pesticides can also have dramatic effects on the environment – poisoning wildlife and contaminating water sources, and passing through the food chain, causing damage along the way. Farmers become trapped on the vicious "pesticides treadmill" – spraying with pesticides destroys the pests' natural enemies, and increases the number of pest outbreaks, as a result of which the farmer uses more pesticides, with the associated negative consequences.

Pesticides are big business, with companies continually consolidating. By the end of 2001 seven agrochemical corporations – labelled the "Gene Giants" because of their investment in biotechnology and promotion of GM crops – controlled over 80 percent of the pesticide market. These companies are expanding their markets in developing countries, despite concerns that the users are ill-equipped to handle hazardous products, and evidence of the destructive impact of pesticides on biodiversity. Another concern is that the planting of GM crops will encourage the use of more agrochemicals. For example, Roundup Ready soybeans are genetically modified to be resistant to the herbicide glyphosate, so that this chemical (which would normally kill the soya) can be blanket sprayed to kill weeds in and around the crop.

Pesticides sales worldwide

1979–1997
US$ billions

$30.2

$26.4

$11.7

1980 1990 1999

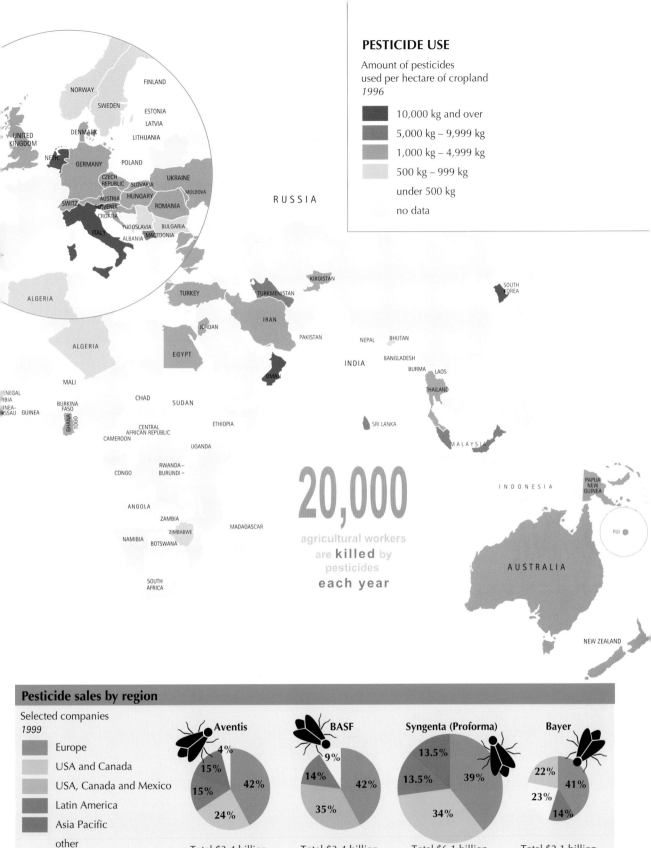

PESTICIDE USE

Amount of pesticides
used per hectare of cropland
1996

- 10,000 kg and over
- 5,000 kg – 9,999 kg
- 1,000 kg – 4,999 kg
- 500 kg – 999 kg
- under 500 kg
- no data

20,000

agricultural workers
are **killed** by
pesticides
each year

Pesticide sales by region

Selected companies
1999

- Europe
- USA and Canada
- USA, Canada and Mexico
- Latin America
- Asia Pacific
- other

Aventis
4%
15%
15%
24%
42%
Total $3.4 billion

BASF
9%
14%
35%
42%
Total $3.4 billion

Syngenta (Proforma)
13.5%
13.5%
34%
39%
Total $6.1 billion

Bayer
22%
23%
14%
41%
Total $2.1 billion

Declining importance

Agricultural labor force as percentage of world labor force
1950–2000

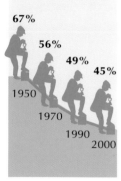

67%

56%

49%

45%

1950

1970

1990

2000

At peak
harvest time
in Kenya

30%

of coffee pickers
are children

THE LAND is the main source of both food and income for just under half the people in the world. The number of people working on the land is increasing, although, as a percentage of the total labor force, the agricultural sector is in decline.

The number of women involved in agriculture also appears to be increasing – although this may be because women's contribution to subsistence farming is becoming more widely recognized. Children, too, play an important part in tending family land, and also make up a sizable proportion of the commercial labor force – around 25 percent of sugar-cane workers in northeastern Brazil, for example. In South Asia and Latin America many children work as bonded laborers to repay family debts.

In developing countries human labor, plus some equipment and animal power, is the energy source for growing food. In parts of Africa this vital supply of energy is being cut off by the devastation brought about by AIDS – a situation that is predicted to have a disastrous affect on food security in many countries.

In industrialized countries, where much of the cultivation and harvesting is done by machinery, the percentage of the labor force employed in agriculture declined dramatically during the 20th century. In 1900 some 50 percent of the French labor force worked on the land. By 2000 it was just over 3 percent. But although few people are working *on* the land, many more are working *for* the land – making fertilizers, agrochemicals, tractors, buildings, pharmaceuticals. The role of this "hidden" agricultural labor force is to increase the productivity of the remaining land workers.

Many countries are turning from labor intensive methods to the use of machines, often in order to produce crops for export. But where the process of replacing human labor with machines has not been undertaken as part of a larger economic plan, it has led to high levels of rural unemployment and the inevitable transformation of agricultural laborers into the urban poor.

Wage levels

Agricultural wage as a percentage of manufacturing wage
1990–1994
selected countries

Romania
157%

Canada
108%

China
75%

Kenya
60%

Australia
58%

Turkey
37%

Israel
30%

Mexico
15%

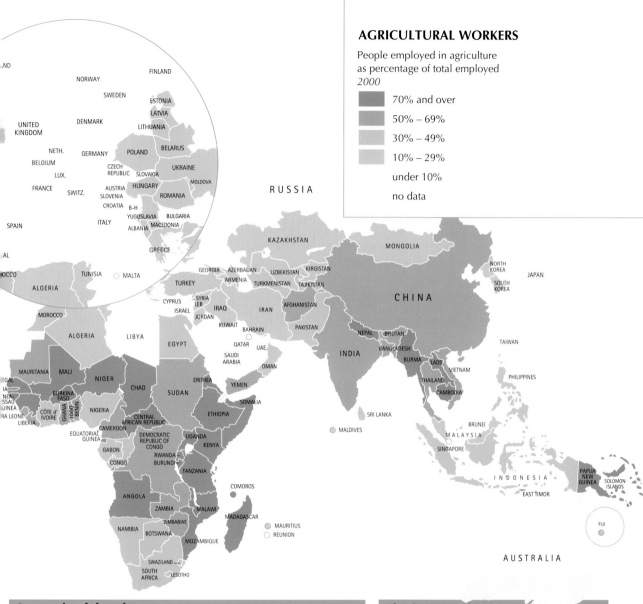

AGRICULTURAL WORKERS

People employed in agriculture
as percentage of total employed
2000

- 70% and over
- 50% – 69%
- 30% – 49%
- 10% – 29%
- under 10%
- no data

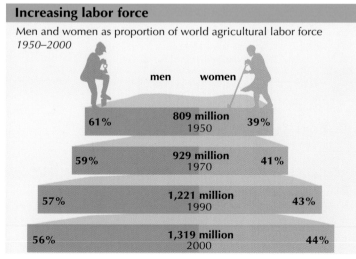

Increasing labor force

Men and women as proportion of world agricultural labor force
1950–2000

men   women

61%	**809 million** 1950	39%
59%	**929 million** 1970	41%
57%	**1,221 million** 1990	43%
56%	**1,319 million** 2000	44%

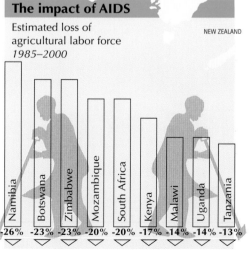

The impact of AIDS

Estimated loss of
agricultural labor force
1985–2000

NEW ZEALAND

Namibia	Botswana	Zimbabwe	Mozambique	South Africa	Kenya	Malawi	Uganda	Tanzania
-26%	-23%	-23%	-20%	-20%	-17%	-14%	-14%	-13%

49

19 | URBAN FARMING

AROUND 800 MILLION city dwellers worldwide – including some in industrialized countries – use their agricultural skills to feed themselves and their families. The world's cities are expanding at an ever-increasing rate. People are leaving agricultural regions no longer able to support them in order to find employment in urban areas. Some of the world's largest cities are now to be found in developing countries. However, most urban immigrants, do not find employment on arrival, but poverty and malnutrition.

As well as growing food for their own consumption, around 200 million also earn a living growing food and rearing livestock to sell at local markets, while a further 150 million are employed as laborers on urban farms. The outskirts of most cities in Africa, Asia and Latin America support thousands of cattle, goats, pigs, chickens and rabbits, and both small and large livestock are also found in inner-city areas. When Hong Kong had an outbreak of avian flu in 1997, over one million chickens, housed in residential areas had to be destroyed – an unforeseen census.

More recently, as a result of the economic slow-down, in particular in East Asia, urban unemployment and poverty has risen. Increased urban food production is in many cases a response to these problems. In times of severe political or economic crisis, when supply lines or currencies collapse, large cities are particularly vulnerable to food shortages.

During Indonesia's financial crisis in 1998, when food prices rose by 70 percent, the government encouraged people in Jakarta to grow food to prevent a breakdown of the fresh-food supply.

Supplying nutritionally adequate and safe food to city dwellers is a substantial challenge for governments and planners. Urban farming creates employment and income, increases food security and can improve the urban environment, but it faces stiff competition for land from developers. In order to ensure a healthy future for urban small-holdings, policy-makers need a properly integrated policy that anticipates the changing needs of both rural and urban populations, and the links between them. Health controls are also important.

$500

million
worth of fruit
and vegetables
is produced
by urban farmers
worldwide

Havana, Cuba
■ 41% of city area
● 58% of Cuba's vegetables

Mexico City, Mexico
▲ 1% employed in agriculture

La Paz, Bolivia
● 30% of agricultural requirements

Montevideo, Uruguay
6% of national pig ● production

Over 20 percent of Dar-es-Salaam, Tanzania is used for farming. Around 20 percent of workers do agricultural work, and 35,000 families in the city and its suburbs depend on income from agriculture. Urban farmers produce 90 percent of the leafy vegetables consumed in the city, and also raise livestock, including 6.5 million chickens.
Some farms are beside rivers and have irrigation systems, but others rely on rainfall.

30,000 people in London rent small plots of land on which they grow vegetables and fruit.

FOOD PRODUCTION IN URBAN AND SUBURBAN AREAS

- ▲ percentage of inhabitants or households involved in agriculture in urban or suburban area

- ■ percentage of city area used for agriculture

- ● percentage of agricultural needs of city met by urban and / or suburban production

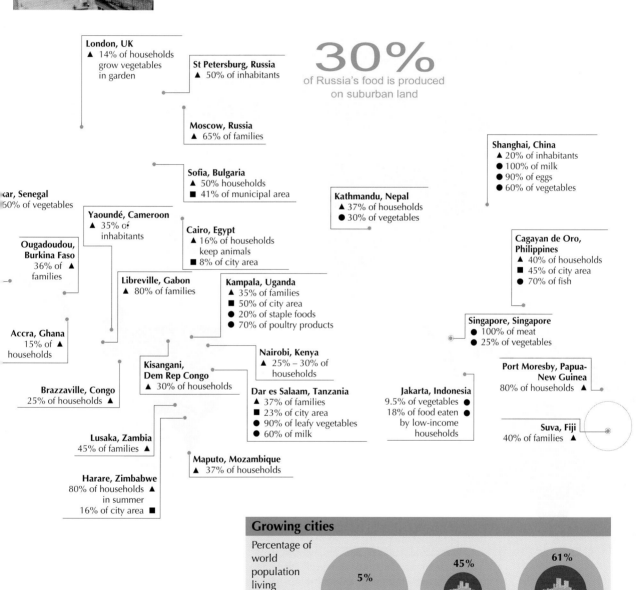

London, UK
▲ 14% of households grow vegetables in garden

St Petersburg, Russia
▲ 50% of inhabitants

30%
of Russia's food is produced on suburban land

Moscow, Russia
▲ 65% of families

Shanghai, China
▲ 20% of inhabitants
● 100% of milk
● 90% of eggs
● 60% of vegetables

Sofia, Bulgaria
▲ 50% households
■ 41% of municipal area

Kathmandu, Nepal
▲ 37% of households
● 30% of vegetables

...kar, Senegal
...0% of vegetables

Yaoundé, Cameroon
▲ 35% of inhabitants

Cairo, Egypt
▲ 16% of households keep animals
■ 8% of city area

Cagayan de Oro, Philippines
▲ 40% of households
■ 45% of city area
● 70% of fish

Ougadoudou, Burkina Faso
36% of ▲ families

Libreville, Gabon
▲ 80% of families

Kampala, Uganda
▲ 35% of families
■ 50% of city area
● 20% of staple foods
● 70% of poultry products

Singapore, Singapore
● 100% of meat
● 25% of vegetables

Accra, Ghana
15% of ▲ households

Nairobi, Kenya
▲ 25% – 30% of households

Port Moresby, Papua-New Guinea
80% of households ▲

Kisangani, Dem Rep Congo
▲ 30% of households

Dar es Salaam, Tanzania
▲ 37% of families
■ 23% of city area
● 90% of leafy vegetables
● 60% of milk

Jakarta, Indonesia
9.5% of vegetables ●
18% of food eaten ●
by low-income households

Brazzaville, Congo
25% of households ▲

Lusaka, Zambia
45% of families ▲

Suva, Fiji
40% of families ▲

Maputo, Mozambique
▲ 37% of households

Harare, Zimbabwe
80% of households ▲ in summer
16% of city area ■

Growing cities

Percentage of world population living in cities *1900–2025 projected*

5%	45%	61%
1900	1990s	2025

Fish consumption

Average amount of
fish eaten
per person per year
1997

- 27.7 kg — industrialized countries
- 25.7 kg — China
- 13.7 kg — rest of Asia
- 7.1 kg — Africa

Where does fish come from?

1999
percentages

**Total:
126.5 million tonnes**

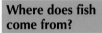
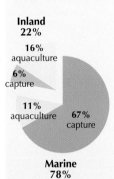

**Inland
22%**

16% aquaculture

6% capture

11% aquaculture

67% capture

**Marine
78%**

Where does it go?

1999
percentages

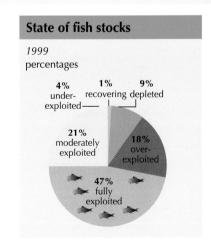

2% other

24% fishmeal, fish oil

74% human consumption

ONE BILLION PEOPLE worldwide rely on fish as their main source of animal protein. The amount consumed varies enormously, depending on availability and income.

Most of the fish we eat is processed – frozen, canned, cured, or turned into animal feed. In 1998 only 36 percent was sold as fresh fish, although this represented an increase compared with the early 1990s.

Most commercial fishing takes place in the world's seas and oceans. Inland waters, mainly freshwater, account for only 6 percent of the world's fish catch, but this valuable source of protein in developing countries is especially vulnerable to over-fishing and to damage by effluent and agricultural chemicals.

The total world fish catch increased by around 6 percent each year throughout the 1950s and 1960s, but this rate of increase slowed to 2 percent a year in the 1970s and 1980s, and the 1990s saw it leveling off. Many of the world's oceans are now being pushed to the limit as an increasing proportion of their fish stocks are exploited at or beyond sustainable levels. National governments and international organizations are working on policies to protect fish stocks, but with no single body controlling fishing in international waters any policy relies on voluntary cooperation from the countries and fishing fleets concerned.

As natural fish stocks decline, the demand for non-native fish in industrialized countries is being met by the rapidly expanding aquaculture industry (fish farming), which now accounts for around a third of fish eaten around the world. Around 90 percent of aquaculture takes place in low-income developing countries. As well as being a valuable export, it is increasingly seen as important to domestic food security. It needs to be carefully controlled, however, since unregulated aquaculture can be damaging to human health, as contaminants in fish feed are passed down the food chain. There are also concerns about environmental pollution.

China alone accounts for almost one third of the world's fish production, and aquaculture is becoming increasingly important. Most of the fish is for domestic consumption, or for use as animal feed.

State of fish stocks

1999
percentages

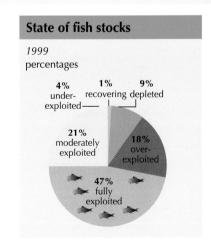

- 4% under-exploited
- 1% recovering
- 9% depleted
- 21% moderately exploited
- 18% over-exploited
- 47% fully exploited

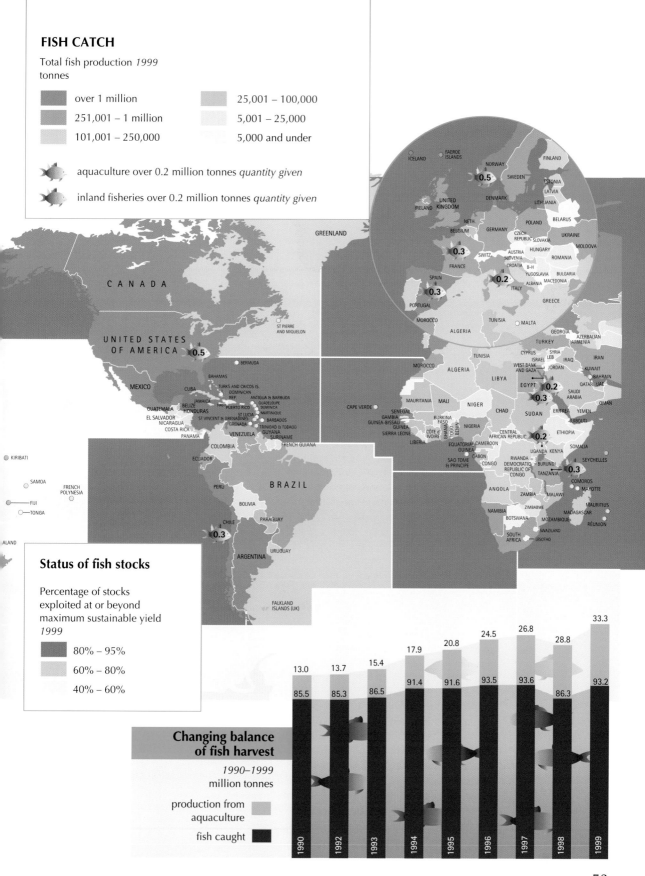

FISH CATCH

Total fish production *1999*
tonnes

- over 1 million
- 251,001 – 1 million
- 101,001 – 250,000
- 25,001 – 100,000
- 5,001 – 25,000
- 5,000 and under

aquaculture over 0.2 million tonnes *quantity given*

inland fisheries over 0.2 million tonnes *quantity given*

Status of fish stocks

Percentage of stocks
exploited at or beyond
maximum sustainable yield
1999

- 80% – 95%
- 60% – 80%
- 40% – 60%

Changing balance of fish harvest

1990–1999
million tonnes

- production from aquaculture
- fish caught

Year	Aquaculture	Fish caught
1990	13.0	85.5
1992	13.7	85.3
1993	15.4	86.5
1994	17.9	91.4
1995	20.8	91.6
1996	24.5	93.5
1997	26.8	93.6
1998	28.8	86.3
1999	33.3	93.2

53

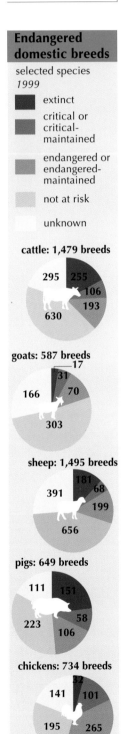

Endangered domestic breeds

selected species
1999

- extinct
- critical or critical-maintained
- endangered or endangered-maintained
- not at risk
- unknown

cattle: 1,479 breeds

295 255 106 193 630

goats: 587 breeds

17 31 70 166 303

sheep: 1,495 breeds

181 68 391 199 656

pigs: 649 breeds

111 151 58 223 106

chickens: 734 breeds

32 141 101 195 265

AGRICULTURAL BIODIVERSITY includes not only the animals and plants used for food, but the diversity of species that support food production – micro-organisms in the soil, pest-predators, crop pollinators – and the wider environment within which the agricultural ecosystem is located.

The genetic diversity of our food has arisen from a 10,000-year process in which wild species have been selected and bred to create the domesticated varieties used today. The regions where these developments took place are defined as centers of diversity of specific crops and related wild species. Such diversity is important because it provides a pool of genes that have developed natural resistance to pests and other environmental stress over time, and will help to ensure the future survival of key food crops. Relying on a single variety of crop to provide food makes a population vulnerable to pests and disease.

But the genetic diversification of food crops and animal breeds is diminishing rapidly. At the beginning of the 21st century it is estimated that only 10 percent of the variety of crops that have been developed in the past are still being farmed, with many local varieties being replaced by a small number of improved varieties, often involving non-native plants. Large numbers of animals are known to have become extinct, and nearly a third of domestic breeds are threatened with extinction. A quarter of the world's fish stocks are being fished above sustainable levels.

Privately and publicly owned gene banks – such as those of the Consultative Group of International Agricultural Research centers (CGIAR) – conserve genetic material artificially. In regions relatively untouched by industrial farming practices, however, a huge variety of crops is still in use. Indigenous farmers in Peru, for example, cultivate 3,000 different varieties of potato, while 5,000 varieties of sweet potato are cultivated in Papua New Guinea. But many areas of genetic diversity in the wild are under threat from "contamination" and domination by introduced varieties, including genetically modified crops.

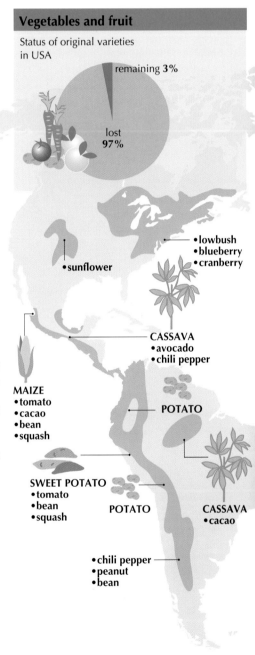

Vegetables and fruit

Status of original varieties in USA

remaining **3%**

lost **97%**

- sunflower
- lowbush
- blueberry
- cranberry

CASSAVA
- avocado
- chili pepper

MAIZE
- tomato
- cacao
- bean
- squash

POTATO

SWEET POTATO
- tomato
- bean
- squash

POTATO

CASSAVA
- cacao

- chili pepper
- peanut
- bean

GENETIC DIVERSITY

Genetic diversity

area in which wild ancestral species
of food crops are found

Relevant food crops given. Most important food crops
(in terms of weight produced) shown as symbols.

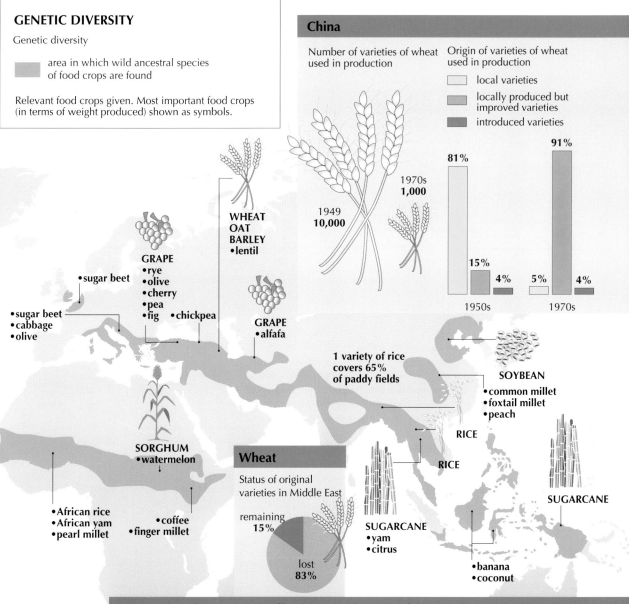

China

Number of varieties of wheat
used in production

Origin of varieties of wheat
used in production

- local varieties
- locally produced but improved varieties
- introduced varieties

81%
91%
15%
4%
5%
4%
1950s
1970s

1949
10,000

1970s
1,000

GRAPE
•rye
•olive
•cherry
•pea
•fig •chickpea

•sugar beet

•sugar beet
•cabbage
•olive

WHEAT
OAT
BARLEY
•lentil

GRAPE
•alfafa

1 variety of rice
covers 65%
of paddy fields

SOYBEAN
•common millet
•foxtail millet
•peach

RICE

RICE

SORGHUM
•watermelon

SUGARCANE

•African rice
•African yam
•pearl millet

•coffee
•finger millet

SUGARCANE
•yam
•citrus

•banana
•coconut

Wheat

Status of original
varieties in Middle East

remaining
15%

lost
83%

International agreements affecting genetic ownership

Ownership of, and access to, agricultural biodiversity are issues of high international politics. Traditionally, agricultural genetic diversity has been a shared resource, and farmers have been free to save seeds to use in future plantings, but a series of agreements from the 1990s has jeopardized this free access to food crops.

1991 Union Internationale pour le protection des Obtentions Végétales (UPOV) recognized breeders' rights and gave legal ownership of industrialized seeds to the companies that developed them.

1992 UN Commission on Environment and Development's World Summit in Rio:
- Trade-Related Intellectual Property Rights agreement extends ownership to living forms
- Convention on Biological Diversity recognized national sovereignty over key genetic resources.

2002 The UN's Food and Agriculture Organization (FAO), agreed an International Treaty on Plant Genetic Resources for Food and Agriculture, allowing common shared access to a limited number of important crop varieties (subject to final approval by the signatories at their respective national levels).

THE GLOBAL TREND towards sustainable practices in agriculture has been termed "the real green revolution", as a contrast to the "green revolution" of the early 1960s, which saw an increase in mechanized, high-input agriculture.

The real green revolution encourages biodiversity, local self-reliance and organic methods. It is partly fuelled by growing consumer demand in the industrialized countries for organic produce, but is also a response to the environmental problems that have developed as a result of the drive to intensify production that has led to the use of excessive pesticides and fertilizers.

Much of the world is already farmed organically. Of the farms in the developing world that register as organic, around 80 percent do not need to change their practices. And while agrochemical companies are encouraging farmers to use chemicals, organic composts can also dramatically increase yields. A project in Brazil, for example, has demonstrated that use of green manures and cover crops can increase yields by 250 percent. Organically enriched soil holds more moisture, enabling vegetables to be grown even during dry periods.

Many of the techniques devised by organic farmers have now been adopted by conventional farmers, including the "community ecology" approach to controlling pests, which recognizes the role played by natural predators.

Certain countries are leading the world in their adoption of organic methods. Cuba, for example, organically produces 65 percent of its rice and nearly 50 percent of its fresh vegetables. Australia and Argentina have the largest areas under organic management, while some European countries have the highest percentage of land certified organic. The production of organic meat, dairy produce and eggs is increasing, especially in the industrialized countries. Even in the USA, land of intensive poultry rearing, there is a significant rise in organically reared poultry.

80%

of farmers in developing countries **do not need to change their methods** to be certified organic

Organically reared hens

Number of laying hens in USA
1992–1997

1992   **44,000**

1997

538,000

CANADA

USA

MEXICO

CUBA

DOMINICAN REPUBLIC

BELIZEs
HONDURAS
GUATEMALA
EL SALVADOR   NICARAGUA

COSTA RICA

TRINIDAD & TOBAGO
VENEZUELA
SURINAME

COLOMBIA

ECUADOR

PERU

BRAZIL

BOLIVIA

PARAGUAY

CHILE

URUGUAY
ARGENTINA

Increase in organic land

1995–2000
square kilometers

in USA

1995
370 sq km

1997
600 sq km

2000
900 sq km

in EU

1995
1,252 sq km

1997
2,154 sq

2000
3,944 sq km

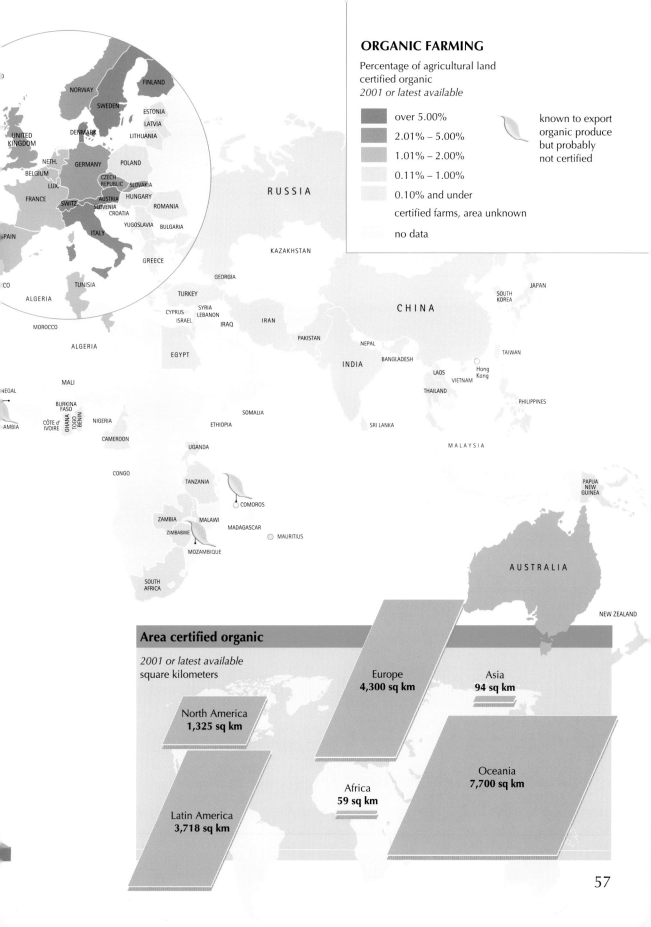

ORGANIC FARMING

Percentage of agricultural land
certified organic
2001 or latest available

- over 5.00%
- 2.01% – 5.00%
- 1.01% – 2.00%
- 0.11% – 1.00%
- 0.10% and under
- certified farms, area unknown
- no data

known to export
organic produce
but probably
not certified

Area certified organic

2001 or latest available
square kilometers

Europe
4,300 sq km

Asia
94 sq km

North America
1,325 sq km

Africa
59 sq km

Oceania
7,700 sq km

Latin America
3,718 sq km

TRADE

"If multinational companies are successful in creating a truly global agricultural system in which they control prices and movement of commodities, the right of each country to establish its own farm policies will be destroyed."
– Jorge Calderon, Professor of Economics, University of Mexico

Food exports

1961 and 1999
million tonnes

774

190

1961   1999

INTERNATIONAL TRADE IN FOOD has expanded significantly in recent decades. Between 1961 and 1999 there was a four-fold increase in the amount of food exported. By 1999 more than 1 in 10 food products was exported, with 580 million tonnes more food traded than in 1961. Large increases in the export of cereals, fruit and vegetables, meat and milk account for much of this expansion.

In 2000, eight of the wealthiest countries accounted for almost half the value of all agricultural exports. The countries of western Europe and North America account for 60 percent of all food exports by value. Both North America and Latin America export more agricultural products than they import – the USA alone accounting for 13 percent of all agricultural *exports* in 2000. Japan, on the other hand, accounted for 11 percent of agricultural *imports* worldwide.

Many countries are now economically dependent on food exports. Agricultural products account for over 80 percent of exports from many Central American countries and over half of all merchandise exported from New Zealand and Côte d'Ivoire.

One of the main drivers of this expansion in international trade in food has been the World Trade Organization (WTO). The WTO's Agreement on Agriculture has promoted trade liberalization through reductions in agricultural subsidies, tariffs and import quotas. The establishment of trading blocks such as the European Union Single Market and the North American Free Trade Agreement has also removed barriers to regional trade, stimulating increased food trade flows.

There will be a further expansion in international trade in food if policies aimed at promoting trade liberalization are implemented. In World Trade Organization (WTO) meetings, the Cairns Group, consisting of the major food producers and exporters, is negotiating for additional reforms that would remove barriers to international trade in food.

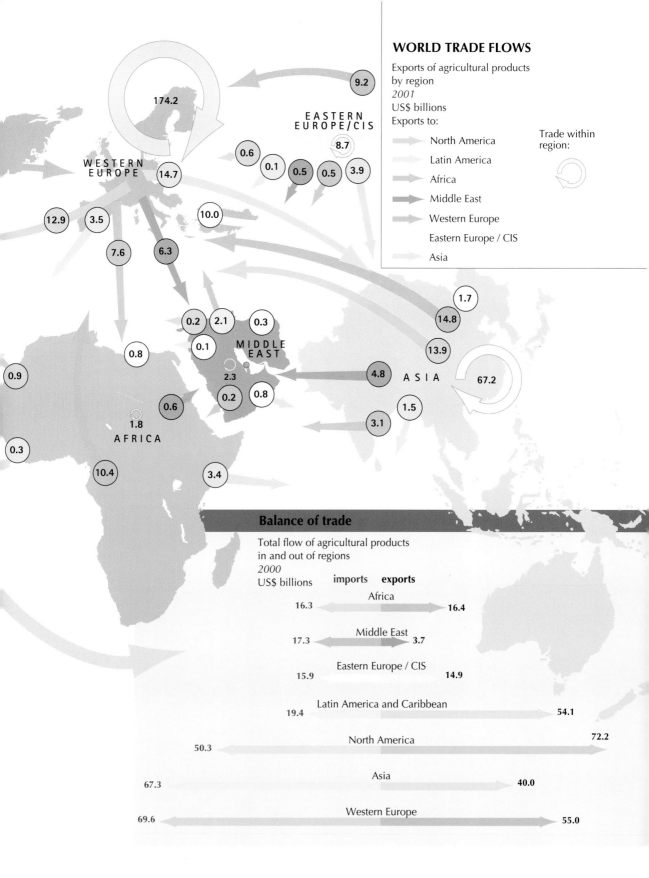

WORLD TRADE FLOWS

Exports of agricultural products
by region
2001
US$ billions
Exports to:

→ North America

→ Latin America

→ Africa

→ Middle East

→ Western Europe

→ Eastern Europe / CIS

→ Asia

Trade within
region:

EASTERN
EUROPE/CIS

9.2

174.2

WESTERN
EUROPE 14.7

0.6 0.1 0.5 0.5 3.9

8.7

12.9 3.5

7.6 6.3

10.0

0.2 2.1 0.3

0.1

MIDDLE
EAST

2.3

0.2 0.8

1.7

14.8

13.9

4.8 ASIA 67.2

1.5

3.1

0.9

0.8

0.6

0.3

1.8

AFRICA

10.4

3.4

Balance of trade

Total flow of agricultural products
in and out of regions
2000
US$ billions

	imports	exports
Africa	16.3	16.4
Middle East	17.3	3.7
Eastern Europe / CIS	15.9	14.9
Latin America and Caribbean	19.4	54.1
North America	50.3	72.2
Asia	67.3	40.0
Western Europe	69.6	55.0

ANIMAL TRANSPORT WORLDWIDE

AROUND 44 MILLION cattle, pigs and sheep are traded across the world each year. Millions more are transported over long distances by road and rail within such countries as Australia and the USA. There are cultural and economic reasons for the trade.

Australia exports over 5 million live sheep each year to the Middle East and over 1 million cattle to Southeast Asia and the Middle East to meet the demand for live animals slaughtered according to halal procedures, under which the animals are usually not stunned before slaughter.

Each year almost 100,000 sheep die of disease and injuries sustained during the long journey to the Middle East, which may involve lengthy travel across Australia to reach a port, a sea crossing of up to three weeks, and an unspecified period of time in a feedlot awaiting slaughter.

The trade has its opponents within Australia, but they are fighting a lucrative business that is not only worth nearly US$127 million a year, but also opens up markets for other goods in these regions.

In North America, animals are bought and sold across borders as if they were just like any other commodity, as traders aim to get the best return on their investment. The USA imports 2 million cattle and 4 million pigs each year, mainly from Canada and Mexico. Within the USA, animals are transported long distances from where they are reared or imported to where they are fattened and slaughtered. Cattle are sent from Montana to Pennsylvania, for example, and weaner pigs born in Ontario are transported to Iowa. Although regulations stipulate a 5-hour rest period every 28 hours, around 1 in 100 cows and cattle arrives at their destination unable to stand.

44
million
cattle
sheep
and
pigs
are traded across
the world
each year

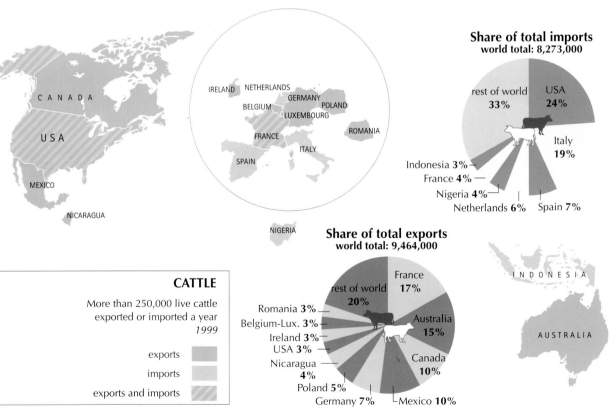

Share of total imports
world total: 8,273,000

- rest of world **33%**
- USA **24%**
- Italy **19%**
- Spain **7%**
- Netherlands **6%**
- Nigeria **4%**
- France **4%**
- Indonesia **3%**

Share of total exports
world total: 9,464,000

- rest of world **20%**
- France **17%**
- Australia **15%**
- Canada **10%**
- Mexico **10%**
- Germany **7%**
- Poland **5%**
- Nicaragua **4%**
- USA **3%**
- Ireland **3%**
- Belgium-Lux. **3%**
- Romania **3%**

CATTLE

More than 250,000 live cattle exported or imported a year
1999

- exports
- imports
- exports and imports

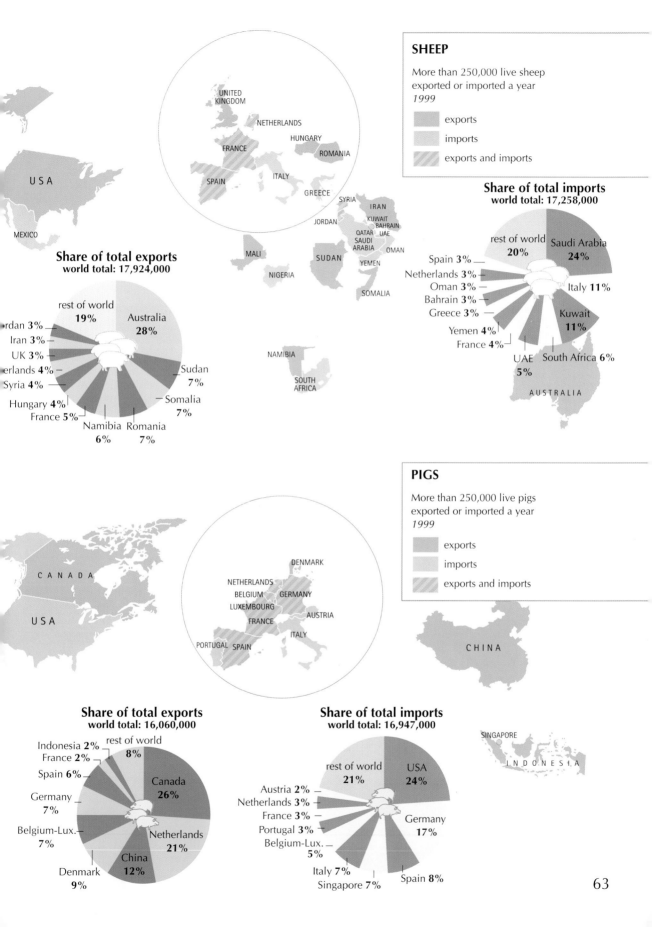

SHEEP

More than 250,000 live sheep
exported or imported a year
1999

- exports
- imports
- exports and imports

UNITED KINGDOM
NETHERLANDS
HUNGARY
FRANCE
ROMANIA
SPAIN
ITALY
GREECE
SYRIA
IRAN
JORDAN
KUWAIT
BAHRAIN
QATAR   UAE
SAUDI ARABIA   OMAN
YEMEN
MALI
SUDAN
NIGERIA
SOMALIA
NAMIBIA
SOUTH AFRICA
AUSTRALIA
USA
MEXICO

Share of total exports
world total: 17,924,000

- Australia **28%**
- rest of world **19%**
- Sudan **7%**
- Somalia **7%**
- Romania **7%**
- Namibia **6%**
- France **5%**
- Hungary **4%**
- Syria **4%**
- erlands **4%**
- UK **3%**
- Iran **3%**
- rdan **3%**

Share of total imports
world total: 17,258,000

- Saudi Arabia **24%**
- rest of world **20%**
- Italy **11%**
- Kuwait **11%**
- South Africa **6%**
- UAE **5%**
- France **4%**
- Yemen **4%**
- Greece **3%**
- Bahrain **3%**
- Oman **3%**
- Netherlands **3%**
- Spain **3%**

PIGS

More than 250,000 live pigs
exported or imported a year
1999

- exports
- imports
- exports and imports

DENMARK
NETHERLANDS
BELGIUM   GERMANY
LUXEMBOURG
FRANCE   AUSTRIA
PORTUGAL   SPAIN   ITALY
CANADA
USA
CHINA
SINGAPORE
INDONESIA

Share of total exports
world total: 16,060,000

- Canada **26%**
- Netherlands **21%**
- China **12%**
- Denmark **9%**
- Belgium-Lux. **7%**
- Germany **7%**
- Spain **6%**
- rest of world **8%**
- Indonesia **2%**
- France **2%**

Share of total imports
world total: 16,947,000

- USA **24%**
- rest of world **21%**
- Germany **17%**
- Spain **8%**
- Italy **7%**
- Singapore **7%**
- Belgium-Lux. **5%**
- Portugal **3%**
- France **3%**
- Netherlands **3%**
- Austria **2%**

63

25 | TRANSPORTING ANIMALS IN EUROPE

EACH YEAR TWO MILLION live pigs, cattle, sheep and horses are transported on long journeys around Europe. Opponents of this trade argue that animals should be slaughtered near to where they were reared, and that the meat should then be transported. This is what happens to 85 percent of UK sheep. The question is why the remaining 15 percent are made to suffer such long, arduous journeys.

It costs four times as much to transport live animals as to transport meat, but animals slaughtered in France are often described as "home-killed" or "home produced", thus attracting a premium price. Moreover, lambing occurs only in certain seasons, and it may be more economic for abattoirs to import animals than for machinery and workers to stand idle.

Animals are also transported for fattening. The relatively strict anti-pollution laws in the Netherlands mean that some of the millions of piglets born there each year have to be fattened in other countries.

Foot and Mouth Disease

The outbreak of Foot and Mouth Disease in the UK in 2001 resulted in over 2,000 cases, and led to 6 million livestock being slaughtered.

It spread over a much wider area than the 1967 outbreak because of the extensive movement of livestock both within the UK and between UK, Ireland and mainland Europe, with an animal passing through a number of hands from birth to slaughterhouse.

After developing in the northeast of England, the disease remained undetected while infected animals were transported via markets to all corners of the country. Once the disease was detected in an abattoir in Essex, steps were taken to prevent animal movement, but by then it was too late.

The role played by the live animal trade in spreading the disease was highlighted when infected British sheep that had been sent to France stopped at the same resting place in northern France as calves shipped from the south coast of Ireland. Foot and Mouth was transmitted from the sheep to the calves, which then took it into the Netherlands.

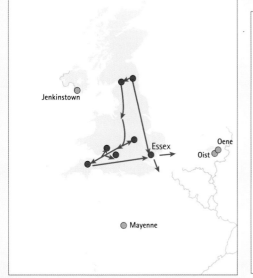

Spread of Foot and Mouth Disease
February – March 2001

← movement of infected animals prior to detection of disease

outbreak of disease outside UK in March 2001

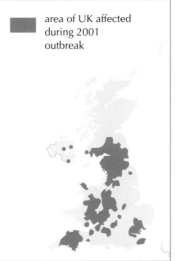

area of UK affected during 2001 outbreak

area of UK affected during 1967 outbreak

Most of the trade is from northern to southern Europe. Many of these journeys take 40 or 50 hours. Some take even longer. Sheep being sent from the UK to Greece, and horses being sent from Lithuania to Sardinia, may travel for over 90 hours.

Overcrowding, high summer temperatures, lack of water and proper ventilation in many vehicles, and the sheer length of the journeys, leads to the animals becoming exhausted, dehydrated and stressed. Some are injured, others collapse. Many die.

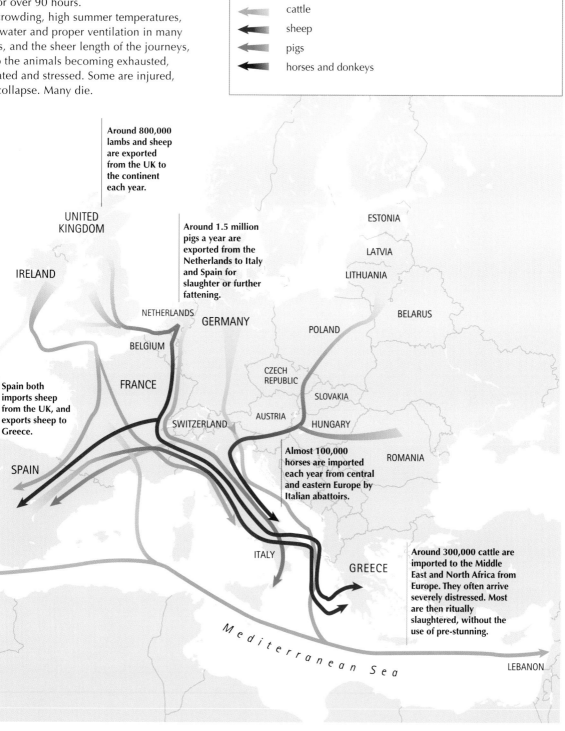

ANIMAL TRANSPORT WITHIN EUROPE

Main routes
2000

→ cattle
→ sheep
→ pigs
→ horses and donkeys

Around 800,000 lambs and sheep are exported from the UK to the continent each year.

Around 1.5 million pigs a year are exported from the Netherlands to Italy and Spain for slaughter or further fattening.

Spain both imports sheep from the UK, and exports sheep to Greece.

Almost 100,000 horses are imported each year from central and eastern Europe by Italian abattoirs.

Around 300,000 cattle are imported to the Middle East and North Africa from Europe. They often arrive severely distressed. Most are then ritually slaughtered, without the use of pre-stunning.

UNITED KINGDOM
IRELAND
NETHERLANDS
GERMANY
BELGIUM
FRANCE
SWITZERLAND
SPAIN
PORTUGAL
ESTONIA
LATVIA
LITHUANIA
BELARUS
POLAND
CZECH REPUBLIC
SLOVAKIA
AUSTRIA
HUNGARY
ROMANIA
ITALY
GREECE
Mediterranean Sea
LEBANON

65

26 FOOD MILES

THE ENVIRONMENTAL IMPACT of the increase in world food trade is felt around the globe. Trade-related transportation is one of the fastest growing sources of greenhouse gas emissions. But because emissions resulting from international air freight and sea freight are not included in national inventories, or in targets under the Kyoto Protocol, there is no incentive for them to be reduced.

The price of sending food by sea fell by over 70 percent between 1980 and 2000, while air freight prices are falling by 3 to 4 percent every year. This does not, however, take into account the actual cost of the damage to the environment and to human health caused by the pollution they create.

The milk trade provides an example of unnecessary food trade. Until recently most people consumed milk produced locally, but from 1961 to 1999 there was a five-fold increase in milk exports, with many countries both importing and exporting large quantities, resulting in millions of extra food miles.

Long-distance movement of food and feed increases the risk of the spread of problems such as Foot and Mouth and Mad Cow Disease (BSE), which can devastate farming communities. It is also a high-risk strategy, dependent on a supply of fuel. Disruption to oil supplies, or sudden increases in fuel prices, could quickly result in food shortages.

A move away from food production for export towards increasing self-sufficiency would benefit many countries. It would improve their food security, and reduce the pollution associated with long-distance transport. However, this is unlikely to happen if international trade in food is further liberalized, and if the environmental costs of distribution are not reflected in the price of food products.

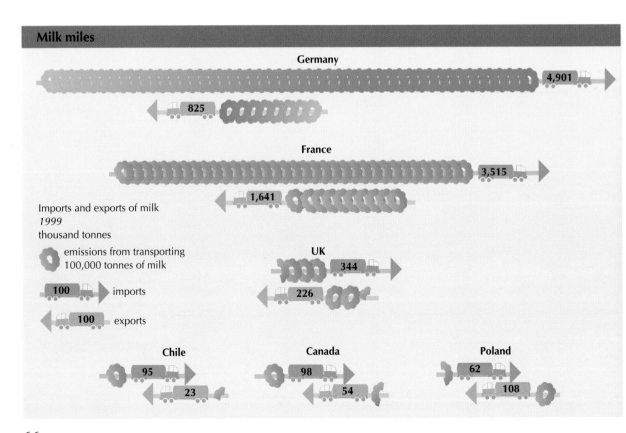

Milk miles

Germany — 4,901 / 825
France — 3,515 / 1,641

Imports and exports of milk
1999
thousand tonnes

emissions from transporting 100,000 tonnes of milk

100 imports

100 exports

UK — 344 / 226

Chile — 95 / 23
Canada — 98 / 54
Poland — 62 / 108

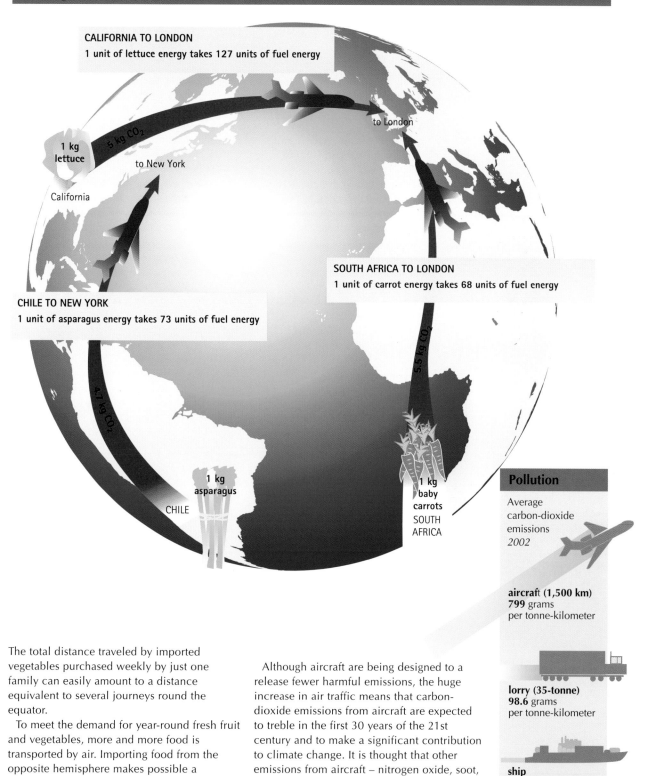

CALIFORNIA TO LONDON
1 unit of lettuce energy takes 127 units of fuel energy

to London

5 kg CO_2

1 kg lettuce

to New York

California

CHILE TO NEW YORK
1 unit of asparagus energy takes 73 units of fuel energy

SOUTH AFRICA TO LONDON
1 unit of carrot energy takes 68 units of fuel energy

5.5 kg CO_2

4.7 kg CO_2

1 kg asparagus

CHILE

1 kg baby carrots
SOUTH AFRICA

Pollution

Average carbon-dioxide emissions
2002

aircraft (1,500 km)
799 grams
per tonne-kilometer

lorry (35-tonne)
98.6 grams
per tonne-kilometer

ship
13 grams
per tonne-kilometer

The total distance traveled by imported vegetables purchased weekly by just one family can easily amount to a distance equivalent to several journeys round the equator.

To meet the demand for year-round fresh fruit and vegetables, more and more food is transported by air. Importing food from the opposite hemisphere makes possible a permanent "dietary summer", but is costly in energy and in environmental damage.

Although aircraft are being designed to release fewer harmful emissions, the huge increase in air traffic means that carbon-dioxide emissions from aircraft are expected to treble in the first 30 years of the 21st century and to make a significant contribution to climate change. It is thought that other emissions from aircraft – nitrogen oxide, soot, water vapor and sulfates – also indirectly enhance the greenhouse effect.

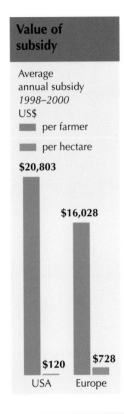

Value of subsidy

Average annual subsidy
1998–2000
US$

■ per farmer

■ per hectare

$20,803

$16,028

$120 $728

USA   Europe

GLOBAL TRADE is not conducted on a level playing field. Governments use various means to support their own agriculture and manufacturing industries by controlling exports and imports. Measures include tariffs (taxes on imported goods), direct subsidies (money paid to the farmer for goods produced), and indirect subsidies (paying a trader to buy and dispose of products that may be surplus to requirement). Export restitution reimburses traders who pay a high price for a domestic product, but sell it for less on the world market.

From the 1940s onward governments used financial incentives to support home production, but since the mid-1970s subsidies have been considered a limitation on free trade. International negotiations resulted in a General Agreement on Tariffs and Trade (GATT), signed by over 100 countries in 1994. It included an Agreement on Agriculture that binds signatories to reducing the subsidies on food paid to farmers and traders. Since 1994 the level of support to farmers in OECD countries has fluctuated but reduced slightly. But in 2002 the US Congress threatened the

GATT consensus by passing a bill increasing spending on agriculture in the USA by 80 percent over a decade.

Developing countries complain that food from high-subsidizing countries is "dumped" in their markets. In 1997 800,000 Mexican farmers faced bankruptcy as a result of direct competition from industrial food production in the USA. Cattle farmers from Burkina Faso to South Africa have been forced out of business due to the dumping of cheap, heavily subsidized meat from the EU.

Those who control the world food trade face complex, long-term challenges – globalization, climate change, bad governance, public health, energy-guzzling supply chains. There is an argument for subsidies, but for programs that provide social and environmental protection. The EU, for instance, introduced an agri-environment program in the early 1990s and is shifting the Common Agricultural Policy from subsidizing farmers to produce food, to subsidizing them to protect and enhance the environment. In the USA up to 7 percent of all support is spent on conservation projects.

SUPPORT TO PRODUCERS

Financial support given to producers as a percentage of total farm receipts averaged across two years
1998–2000
selected countries

■ 60% and over

■ 40% – 49%

■ 20% – 25%

■ 10% – 19%

■ under 10%

■ no data

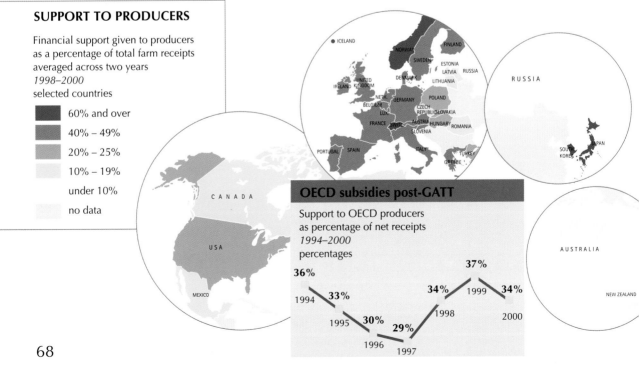

OECD subsidies post-GATT

Support to OECD producers as percentage of net receipts
1994–2000
percentages

36% 1994
33% 1995
30% 1996
29% 1997
34% 1998
37% 1999
34% 2000

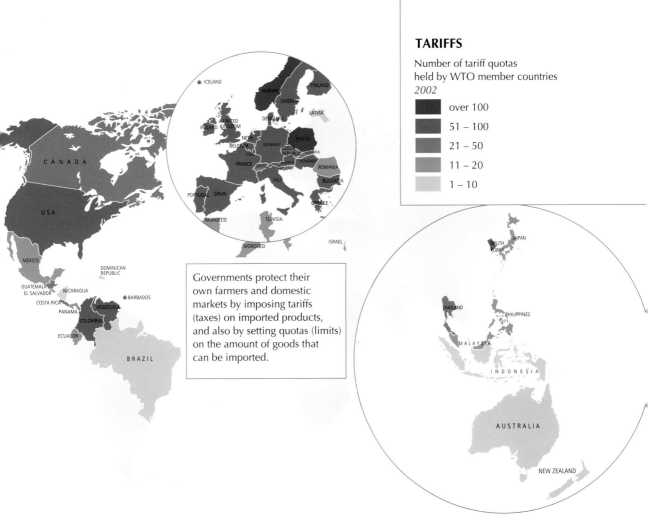

Number of tariff quotas
held by WTO member countries
2002

■	over 100
■	51 – 100
■	21 – 50
■	11 – 20
■	1 – 10

Governments protect their
own farmers and domestic
markets by imposing tariffs
(taxes) on imported products,
and also by setting quotas (limits)
on the amount of goods that
can be imported.

Agricultural support

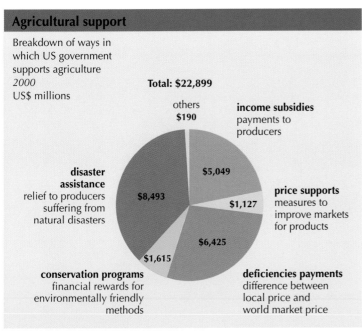

Breakdown of ways in
which US government
supports agriculture
2000
US$ millions

Total: $22,899

others
$190

income subsidies
payments to
producers

$5,049

**disaster
assistance**
relief to producers
suffering from
natural disasters

$8,493

price supports
measures to
improve markets
for products

$1,127

$6,425

$1,615

conservation programs
financial rewards for
environmentally friendly
methods

deficiencies payments
difference between
local price and
world market price

UK direct payments to agriculture

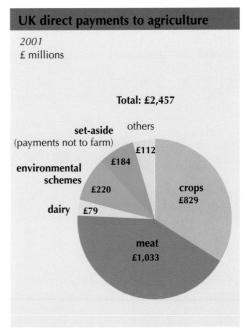

2001
£ millions

Total: £2,457

set-aside
(payments not to farm)
£184

others
£112

**environmental
schemes**
£220

dairy **£79**

**crops
£829**

**meat
£1,033**

69

28 | TRADE DISPUTES

What's the damage?

Estimates of amount owed by USA following 2002 WTO ruling on the illegality of tax exemptions

EU estimate: $4 billion

US estimate: $1 billion

THE BATTLE to secure markets for the increasing amount of food traded around the world frequently results in trade disputes. Most of these are settled between the countries or trade blocs involved. In some cases, however, a dispute settlement ruling is sought from the World Trade Organization (WTO).

Some trade disputes arise because one country is accused of subsidizing agricultural trade in ways that violate WTO Agreement on Agriculture, enabling the subsidized product to be sold at a price below that for a like item in the target market. In 2002 the European Union (EU) won a dispute with the USA over tax exemptions to US companies that enable agribusiness firms to undercut European producers. However, the two sides to the dispute came up with widely differing estimates as to the damages that should be paid by the USA.

Disputes also arise when one country or trading bloc establishes a licensing system, under which they allow the duty-free import of goods up to a certain value from another country or trading bloc. The issue of trade licenses, or quotas, is one that has been addressed under the General Agreement on Trade in Services.

In recent years, several trade disputes have arisen when one country or trading bloc refuses to allow the import of the food produced by another on grounds of public health and safety. In such cases the WTO dispute settlement panels are required to refer to standards for food safety established by the Codex Alimentarius Commission, founded in 1962 by members of the United Nations Food and Agriculture Organization and the World Health Organization.

Health and safety

Several disputes have arisen on the grounds of food safety, possible health risks to the consumer, and risk of damage to the environment.

Lindane
The Canadian government passed a ruling in 1999 refusing to allow the import of seed treated with lindane – a pesticide used to protect canola/rapeseed. Lindane manufacturers Crompton Corp are seeking damages of around $100 million under the North American Free Trade Agreement, which allows corporations to sue for loss of anticipated sales and costs associated with complying with government regulations.

Hormones in beef
The EU has, since 1989, banned the import of US beef containing growth hormones – a decision challenged by the USA in 1997 at a WTO dispute resolution panel. The panel ruled in favor of the USA, citing Codex standards. The EU continues to test the hormone to determine whether it can harm human health, while accepting the tariff sanctions resulting from the ruling. The value of the trade lost to the USA is hotly disputed.

Genetically modified foods
The EU proposes legislation requiring the labeling of food products with one percent of genetically modified ingredients. This would adversely affect the import of some foods from the USA, where there is no system for tracing and labelling GM products. Approval for the legislation is expected in 2003, when Codex standards and guidelines for GM foods are issued. The USA claims the legislation would violate WTO rules and would result in $4 billion dollars per year of lost corn and soy exports. The EU estimate is much lower.

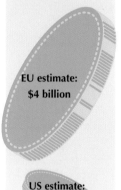

US estimate of value of lost beef trade: $4 billion

GM labeling dispute

EU estimate of value of lost beef trade: $2.2 billion

The banana trade is a multi-million-dollar business, and around a third of all banana exports go to the European Union (EU). A major trade dispute arose in 1993 when the EU tried to reserve 18 percent of its banana market for African and Caribbean countries (members of the ACP trade group), on the basis of historic trade links. The USA claimed that this discriminated against non-ACP banana producers, and against providers of refrigeration and transportation services. The US-owned firms Chiquita and Dole own huge banana plantations in Latin America (the "dollar banana" countries) and control 50 percent of the banana trade.

The dispute was taken to the WTO Disputes Settlement Board, and led to a series of rulings and counter-proposals. In April 1999, the USA imposed $191 million of import duties on specified exports from EU member states, damaging many European small businesses.

In April 2001 a two-stage arrangement was agreed. From 2001 until 2005 a combined tariff-quota system is to operate. A projected 63 percent of imports will attract a comparatively small tariff that the "dollar banana" producers, with their much lower production costs, can easily pay and still price their bananas competitively. A much higher tariff will be payable on remaining imports. The ACP countries will be exempt from all tariffs, effectively leaving them with a projected 18 percent of the trade, with 19 percent for EU internal producers.

But from the beginning of 2006 a tariff-only system will operate, under which the ACP countries will get no preferential treatment. It is recognized that they, and small independent producers in particular, will lose trade, and the interim arrangement is designed to give them time to develop new export products. The value of banana exports from the Windward Islands fell by 58 percent between 1990 and 2000, and radical changes are needed if they are to compete with the big producers.

Banana world trade shares 1999

- Del Monte 8%
- Fyffes 8%
- Noboa 8%
- Dole Food Co 25%
- Chiquita 26%
- other 25%

Total sales of Chiquita US$2,566 million

ACP total export revenue US$262 million

Major banana exporting countries

- countries from which small-scale producers mainly export to EU

- "Dollar banana" countries where production and trade is largely controlled by US, UK, French and Japanese multinational corporations

MEXICO
JAMAICA
DOMINICAN REPUBLIC
BELIZE
GUATEMALA
HONDURAS
NICARAGUA
COSTA RICA
PANAMA
VENEZUELA
SURINAME
COLOMBIA
ECUADOR
WINDWARD ISLANDS
CAPE VERDE
CÔTE d'IVOIRE
GHANA
CAMEROON
SOMALIA
PHILIPPINES

Windward Islands exports

Decreasing value of banana exports from the Windward Islands 1990–2000 US$ million

- $387 1990
- $327 1991
- $376 1992
- $270 1993
- $216 1994
- $240 1995
- $224 1996
- $164 1997
- $183 1998
- $177 1999
- $161 2000

71

Extra national
income per person
from increased
exports in 1990s:

rich countries

**low-income
countries**

THE DISTANCE between producer and consumer has grown ever wider as world trade in food has increased. Food-supply chains have become longer and more complicated, and consumers are often unaware of the source of the food they eat, the conditions for the workers who produce the food, and the environmental impact of its production, processing, packaging and distribution.

In the last few decades an increasing amount of food has been traded worldwide. Trade in cereals has increased by 251 percent, for example, and vegetable oil by a massive 1,233 percent. These increases are not, though, matched by increases in production – it is just that a greater proportion of the food produced is circulating on international markets.

The liberalization of trade following the Uruguay Round of the General Agreement on Tariffs and Trade in 1994 has meant that the world's poorer countries have been forced to open up their markets to foreign imports, while rich countries (notably Europe and the USA) keep their markets more protected. Many have also been encouraged to maximize their foreign earnings by increasing their exports in order to pay off international debts. Land that was used to grow food for local consumption has been turned over to "cash crops" such as coffee, tea, cocoa and horticultural products. This has led to countries becoming dependent on just a few crops for their foreign income. Ethiopia and Burundi, for example, depend on coffee for 60 to 80 percent of their earnings.

Individual farmers, as well as entire economies, are extremely vulnerable to a drop in commodity prices on the world markets. They do not have any reserves to tide them over a bad patch, are often forced to sell their crop at less than cost price, and lose their livelihoods as a result. A whole range of foods, including cocoa, coffee, sugar and fishmeal, are now trading at lower prices in real terms than in 1980. Coffee, tea and cocoa have been worst hit in recent years.

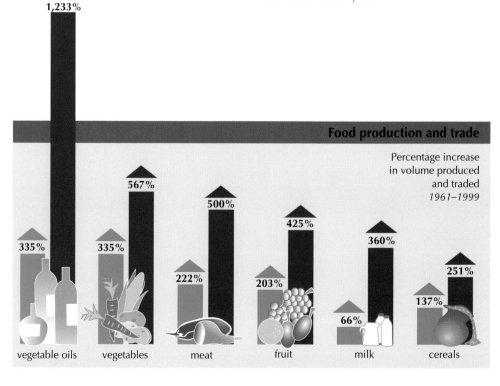

Food production and trade

Percentage increase
in volume produced
and traded
1961–1999

1,233%

567%

500%

425%

360%

335%   335%

251%

222%   203%

137%

66%

vegetable oils   vegetables   meat   fruit   milk   cereals

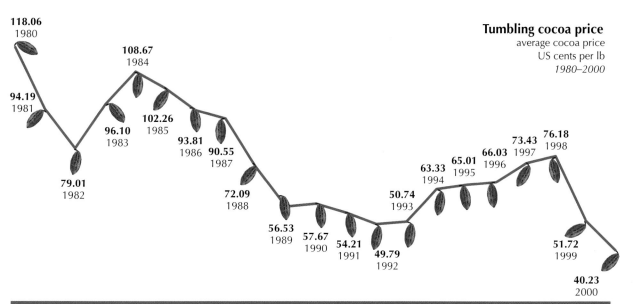

Tumbling cocoa price
average cocoa price
US cents per lb
1980–2000

118.06 1980
94.19 1981
79.01 1982
108.67 1984
96.10 1983
102.26 1985
93.81 1986
90.55 1987
72.09 1988
56.53 1989
57.67 1990
54.21 1991
49.79 1992
50.74 1993
63.33 1994
65.01 1995
66.03 1996
73.43 1997
76.18 1998
51.72 1999
40.23 2000

More trade, less food?

Increasing the amount of food produced for export not only makes a country dependent on world prices for its income, but can adversely affect its domestic food supply. While exports of rice from Thailand more than doubled between 1979 and 1999, domestic rice consumption fell.

Similarly, although vegetable production in Kenya more than doubled between 1969 and 1999, there was a six-fold increase in exports, resulting in a decrease in vegetable consumption among the Kenyans. The largest importer of Kenyan horticultural exports is the UK. Between 70 percent and 90 percent of fresh produce imported to the UK from Africa is controlled by the supermarkets, who pocket around 45 percent of the retail price, while the farmers receive about 17 percent.

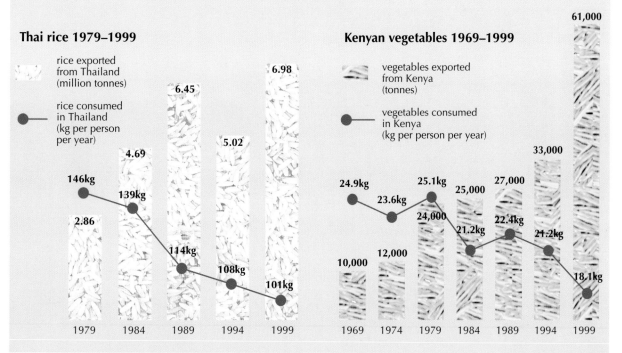

Thai rice 1979–1999

rice exported from Thailand (million tonnes)

rice consumed in Thailand (kg per person per year)

Year	Exported	Consumed
1979	2.86	146kg
1984	4.69	139kg
1989	6.45	114kg
1994	5.02	108kg
1999	6.98	101kg

Kenyan vegetables 1969–1999

vegetables exported from Kenya (tonnes)

vegetables consumed in Kenya (kg per person per year)

Year	Exported	Consumed
1969	10,000	24.9kg
1974	12,000	23.6kg
1979	24,000	25.1kg
1984	25,000	21.2kg
1989	27,000	22.4kg
1994	33,000	21.2kg
1999	61,000	18.1kg

Fairtrade-labeled bananas

Sales worldwide
1998–2001
thousand tonnes

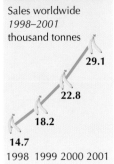

29.1

22.8

18.2

14.7

1998 1999 2000 2001

Who will buy?

Percentage of
25–45 year olds
in the USA
willing to buy
products that benefit
a cause they support
2002

78%

54%

likely
to buy

prepared
to pay
more

THE MODERN MOVEMENT for fair trade originated in the Netherlands in 1989. It aims to create direct and long-term trading links with producers in developing countries, and to ensure that they receive a guaranteed price for their product, on favorable financial terms.

Many small producers outside the "Fairtrade" system, as it is known, do not receive enough money for their product even to cover their costs. Machinery, fertilizers and pesticides, imported from the industrialized world, have become more expensive, while the price of primary commodities such as tea, coffee and sugar on world markets has remained static in real terms for the last 40 years.

Coffee was the first product to be traded fairly. Small coffee producers are badly affected by the volatile international coffee market. Those within the Fairtrade movement are able to cut out local traders and processers, and sell direct to Fairtrade, which pays a price that covers their costs, and extends credit against part of the value of the contract.

The retail price of Fairtrade goods, which may be higher than comparable products, includes a "social premium" that provides funds for investment in social and environmental improvements in the area from which they originate. In the Volta River Estates in Ghana the social premium pays for crops to be weeded by hand, thereby avoiding the use of harmful herbicides. The workers are paid 60 percent above the minimum salary rate, and healthcare is provided on site. The Coopetrabasur in Costa Rica has stopped using the herbicide paraquat and started recycling its plastic waste. The social premium has enabled wages to be raised, environmental specialists to be employed, and housing to be repaired.

Although the Fairtrade movement now represents an estimated $235 million retail sales in Europe alone, it still represents only a small percentage of total trade. Clearly there is scope for further expansion.

Market share

Fair trade sales
as a percentage
of total sales
2000 or latest available

■ coffee
■ tea

3.5% **4%**

Switzerland

1% **2%**

Germany

2.5% **2%**

Denmark

2.5%

0.5%

UK

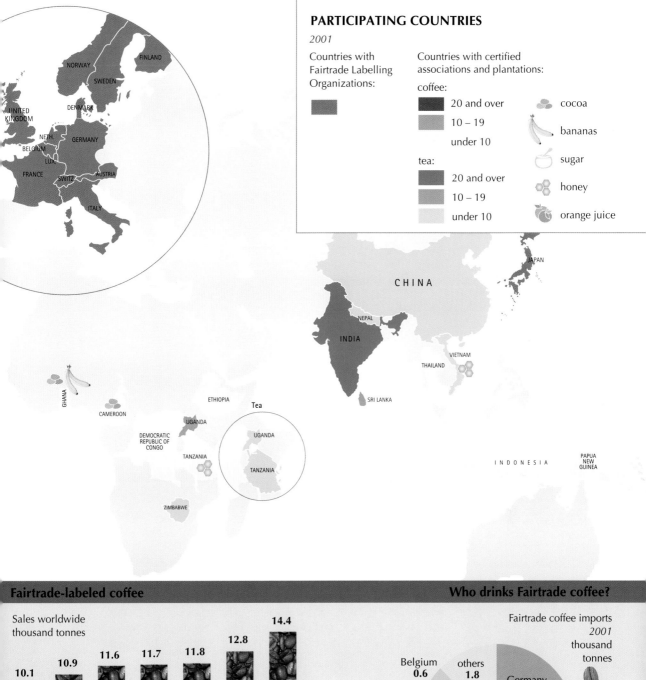

Countries with Fairtrade Labelling Organizations:

Countries with certified associations and plantations:

coffee:
- 20 and over
- 10 – 19
- under 10

tea:
- 20 and over
- 10 – 19
- under 10

- cocoa
- bananas
- sugar
- honey
- orange juice

NORWAY
FINLAND
SWEDEN
UNITED KINGDOM
DENMARK
NETH.
GERMANY
BELGIUM
LUX.
FRANCE
SWITZ.
AUSTRIA
ITALY

JAPAN
CHINA
NEPAL
INDIA
VIETNAM
THAILAND
SRI LANKA

GHANA
CAMEROON
ETHIOPIA
UGANDA
DEMOCRATIC REPUBLIC OF CONGO
TANZANIA
ZIMBABWE

Tea
UGANDA
TANZANIA

INDONESIA
PAPUA NEW GUINEA

Fairtrade-labeled coffee

Sales worldwide thousand tonnes

Year	Sales
1995	10.1
1996	10.9
1997	11.6
1998	11.7
1999	11.8
2000	12.8
2001	14.4

Who drinks Fairtrade coffee?

Fairtrade coffee imports
2001
thousand tonnes

- Germany 3.1
- Netherlands 3.1
- UK 1.6
- Switzerland 1.3
- USA 1.3
- France 0.9
- Denmark 0.7
- Belgium 0.6
- others 1.8

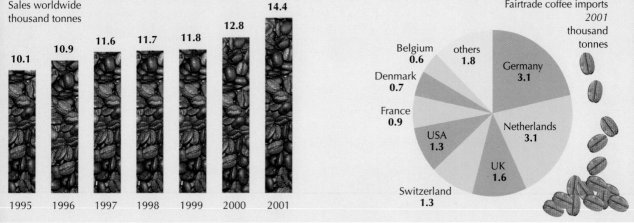

PROCESSING, RETAILING AND CONSUMPTION

4

"We've detached ourselves from our food; it is no longer personal. We want our meals faster and easier. We promote convenience and speed, equating it with quality and value. We are willing to spend billions on diets, nutritional supplements and exercise gyms instead of simply eating right."
– Mas Masumoto, California family farmer, *USA Today*, December 9, 1992

Daily calories

Percentage of calories from each food group
1999

- cereals
- starchy roots
- pulses
- sugars, sweeteners
- total fat
- fruit, vegetables
- meat
- dairy products
- alcohol
- other

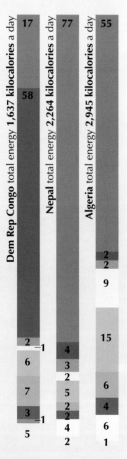

Dem Rep Congo total energy 1,637 kilocalories a day: 17, 58, 2, 1, 6, 7, 3, 1, 5

Nepal total energy 2,264 kilocalories a day: 77, 4, 3, 2, 5, 2, 2, 4, 2

Algeria total energy 2,945 kilocalories a day: 55, 2, 2, 9, 15, 6, 4, 6, 1

A STAPLE FOOD is one that is eaten regularly and which provides a large proportion of a population's energy and nutrients. Cereals are eaten in almost every country of the world, and rice, maize and wheat are the staple food of around 4 billion people (two-thirds of the world's population). Other cereals, such as millet, sorghum and rye are also important.

In Nepal – a low-income developing country – cereals supply over 70 percent of the diet. Other foods, such as sugars, fats and food derived from animals, barely feature. A similar profile is seen in many other low-income countries, particularly those in Southeast Asia. Cereals also make up a high proportion of the diet in North Africa, although here more fat is consumed.

The climate of much of tropical Africa and parts of South America and Oceania is unsuitable for growing most cereals, and here roots and tubers are important. They are an important staple food for over 1 billion people. Cassava (also known as manioc) is the staple food of around 500 million people. It can be processed to make many different local foods, such tapioca, gari, fufu and farinha. The plant originated in South America, but is now widely eaten in Sub-Saharan Africa. Yams, cocoyam, taro, and sweet potato, are all types of root and tuber.

Such food tends to be high in carbohydrates, so it is often known as a starchy staple. Bananas and plantain are also important starchy staples eaten in some tropical countries. But although starchy staples provide carbohydrates they are low in protein. They need to be supplemented by nuts and pulses or under-nutrition is likely to result.

CANADA

USA

MEXICO
BAHAMAS
CUBA
DOMINICAN REPUBLIC
JAMAICA
BELIZE HAITI
GUATEMALA HONDURAS
EL SALVADOR NICARAGUA
COSTA RICA
TRINIDAD & TOBAGO
VENEZUELA GUYANA
PANAMA SURINAME
FRENCH GUIANA
COLOMBIA
ECUADOR
PERU
BRAZIL
BOLIVIA
PARAGUAY
CHILE
URUGUAY
ARGENTINA

Roots and tubers

Manioc being cooked in a market.

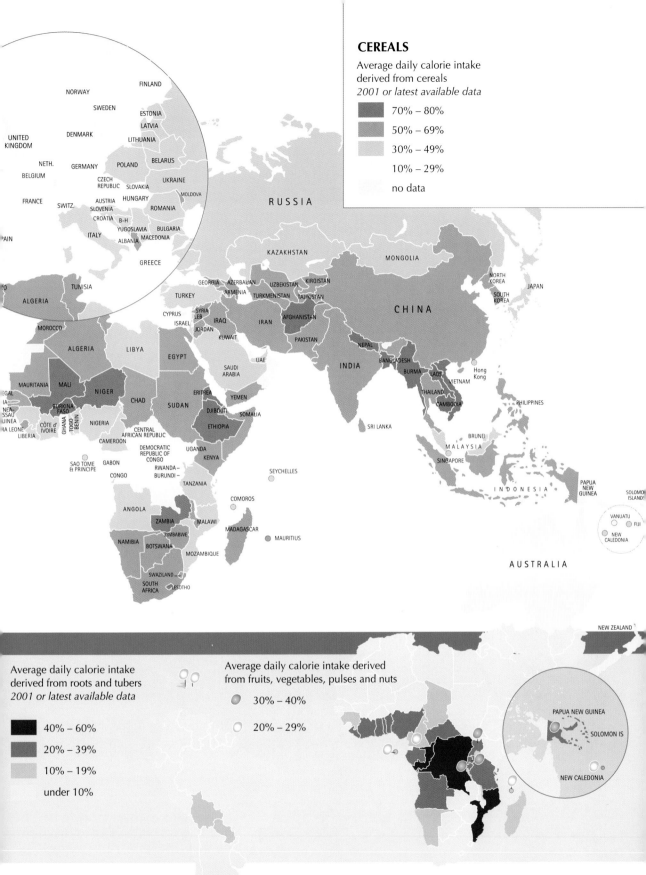

CEREALS

Average daily calorie intake derived from cereals
2001 or latest available data

- 70% – 80%
- 50% – 69%
- 30% – 49%
- 10% – 29%
- no data

Average daily calorie intake derived from roots and tubers
2001 or latest available data

- 40% – 60%
- 20% – 39%
- 10% – 19%
- under 10%

Average daily calorie intake derived from fruits, vegetables, pulses and nuts

- 30% – 40%
- 20% – 29%

DIETS VARY WIDELY around the world and have evolved over many millennia, largely influenced by environmental factors such as climate and ecology. Social factors are also important, with diet partly dependent on whether a social group is agricultural or migratory. Economic conditions, the development of technologies and opportunities for trade have all played a part in influencing diet. Although many traditional diets in developing countries are dominated by a single staple food, this has long since ceased to be the case in industrialized societies, where the variety of food on offer has never been greater.

Diets in industrialized countries generally contain more food of animal origin and less food of plant origin than diets in developing countries. In terms of nutrients, they contain more protein, more fat (including saturated fat) and more sugar, a higher energy density, and relatively little dietary fiber, carbohydrates and antioxidants. In the USA, for example, only about 20 percent of the total dietary energy supply comes from cereals, with almost as much being supplied by sugar and sweeteners, and by fat. A similar dietary pattern is seen in Canada, western Europe and Australasia. The diet of some southern European countries, such as Spain, differs from this pattern in that it contains more fruits and vegetables and its main fat source is vegetable oil, notably olive oil. This has been called "The Mediterranean diet".

Although a shortage of even staple foods continues to be a problem for sectors of the population in many countries, an expansion in food trade, improvements in global communications, and the penetration of new markets by food corporations, are all having an effect on diet. A "nutrition transition" is affecting urban populations in many developing countries, as well as the previously distinctive national diets of countries such as Brazil and Japan.

Daily calories

Percentage of calories from each food group
1999

 cereals

 starchy roots

 pulses

sugars, sweeteners

 total fat

 fruit, vegetables

meat

dairy products

alcohol

other

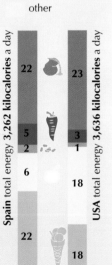

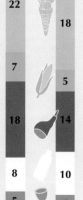

Spain total energy 3,262 kilocalories a day

22
5
2
6
22
7
18
8
5
3

USA total energy 3,636 kilocalories a day

23
3
1
18
18
5
14
10
4
3

CANADA

USA

MEXICO

BAHAMAS

CUBA
DOMINICAN REPUBLIC
JAMAICA
BELIZE   HAITI
HONDURAS
GUATEMALA
EL SALVADOR   NICARAGUA

COSTA RICA
PANAMA
VENEZUELA

TRINIDAD & TOBAGO
GUYANA
SURINAME
FRENCH GUIANA

COLOMBIA

ECUADOR

PERU

BRAZIL

BOLIVIA

PARAGUAY

CHILE

URUGUAY

ARGENTINA

The nutrition transition

The nutrition transition is characterized by:
• a decline in the consumption of traditional staple foods and other traditional food crops, such as pulses and oilseeds
• an increase in intakes of fat, sugar, salt and often animal foods
• an increase in alcohol consumption in non-Islamic countries
• an increase in the consumption of refined and processed foods
• an overall reduction in dietary diversity
Such changes in diet have an impact on health, leading to an increase in diet-related diseases, such as late-onset diabetes, some cancers and cardiovascular disease.

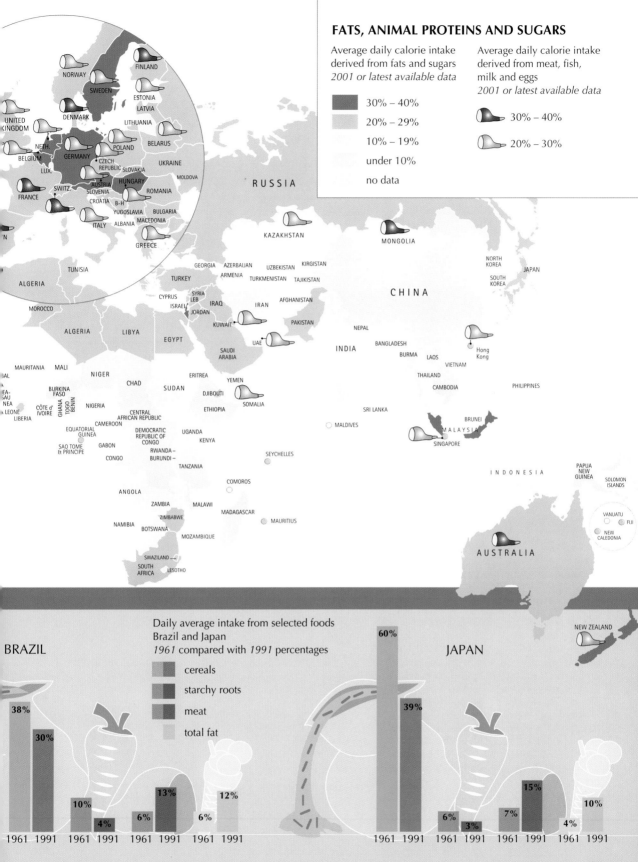

FATS, ANIMAL PROTEINS AND SUGARS

Average daily calorie intake derived from fats and sugars
2001 or latest available data

- 30% – 40%
- 20% – 29%
- 10% – 19%
- under 10%
- no data

Average daily calorie intake derived from meat, fish, milk and eggs
2001 or latest available data

- 30% – 40%
- 20% – 30%

Daily average intake from selected foods
Brazil and Japan
1961 compared with *1991* percentages

- cereals
- starchy roots
- meat
- total fat

33 PROCESSING GIANTS

Top ten

Turnover of top ten global food companies
2000
US$ bn

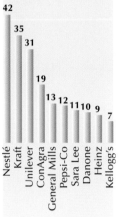

	US$ bn
Nestlé	42
Kraft	35
Unilever	31
ConAgra	19
General Mills	13
Pepsi-Co	12
Sara Lee	11
Danone	10
Heinz	9
Kellogg's	7

FOOD PROCESSING is as old as cooking. With industrialization in the 19th century, specialist processors emerged that took advantage of the scale offered by mass markets and new technologies. New methods of milling, cooking, preserving and transporting foods revolutionized baking, pie making, dairy products, confectionery and soft drinks. Foods such as white bread, which only the rich had previously been able to afford, were made available to mass consumers by the discovery of steel roller mills. Cheap white flour also meant the processor could make a lucrative side business selling the bran and wheat germ (rich in nutrients) as animal food.

Food processors ceaselessly look for new products and new niche markets. Processed food appeals on many levels. It saves time because the cooking and preparation are done in the factory. In the USA, 90 percent of the chicken bought has already been jointed or otherwise processed. Even salads are now offered washed and ready mixed. Processors often claim that their products offer health benefits. In the USA the breakfast cereals market developed first as health foods. Coca Cola, the world's highest value brand, started as a health drink. "Functional foods" are a rapidly expanding market (see pages 90–91) and the suggestion has even been made that the way to combat malnutrition in developing countries is with processed foods fortified with vitamins and minerals.

In most countries, a handful of large food processing companies have emerged. In the USA, the top four processing companies in each line of business between them control a large share of the market – around 80 percent in the case of beef and soybean. And their control does not begin at the point at which the animal is killed, or the crop harvested. Increasingly, the processors are reaching back down the food chain, not necessarily owning but controlling every aspect of production. A farmer wishing to supply one of the large processors becomes totally dependent on the company, with devastating consequences should the processor decide not to renew the contract.

At the start of the 21st century there is a two-tier structure in food processing. A number of international giants exist alongside hundreds of thousands of small local or national firms. Such consolidation is not, however, limited to the processing corporations. Retailers have also merged to form huge and powerful companies, and the food processing companies are increasingly finding that it is the supermarkets that are dictating the terms on which business is done.

Top world brands

Brands are both molders and reflectors of mass consciousness. Food processors, who may own several different brands, invest heavily in advertising their brands so that they are instantly recognizable and desirable. This has become even more important as the retailers have increased in power.

Local brands compete to enter the global brand race. In Europe, by the early 1990s, an estimated 10,000 new food products were coming on to the market each year, with only 10 percent surviving the year. In Canada between 1986 and 1991, the number of products coming to market doubled.

Value of food and drink brands
2000
US$ bn

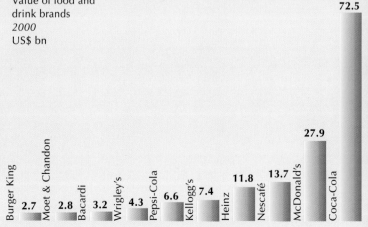

Brand	US$ bn
Burger King	2.7
Moet & Chandon	2.8
Bacardi	3.2
Wrigley's	4.3
Pepsi-Cola	6.6
Kellogg's	7.4
Heinz	11.8
Nescafé	13.7
McDonald's	27.9
Coca-Cola	72.5

Concentration ratio

■ US market share controlled by top four companies *2001 or latest available data*

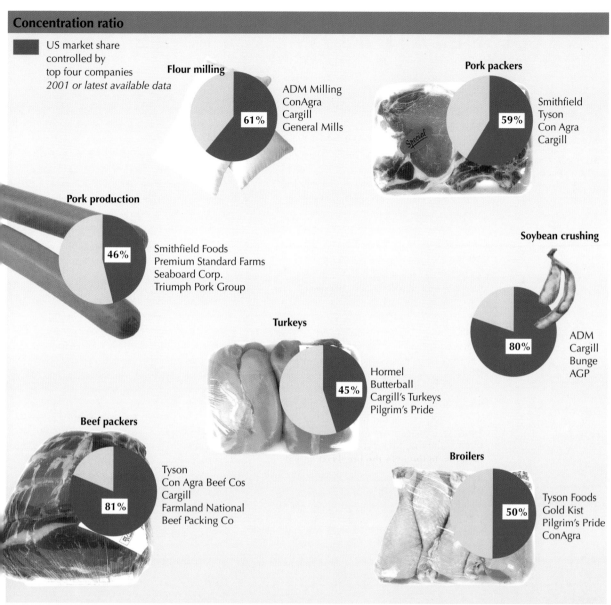

Flour milling
61%

ADM Milling
ConAgra
Cargill
General Mills

Pork packers
59%

Smithfield
Tyson
Con Agra
Cargill

Pork production
46%

Smithfield Foods
Premium Standard Farms
Seaboard Corp.
Triumph Pork Group

Soybean crushing
80%

ADM
Cargill
Bunge
AGP

Turkeys
45%

Hormel
Butterball
Cargill's Turkeys
Pilgrim's Pride

Beef packers
81%

Tyson
Con Agra Beef Cos
Cargill
Farmland National
Beef Packing Co

Broilers
50%

Tyson Foods
Gold Kist
Pilgrim's Pride
ConAgra

Increasing control

Change in percentage of business controlled by top four US companies
2000

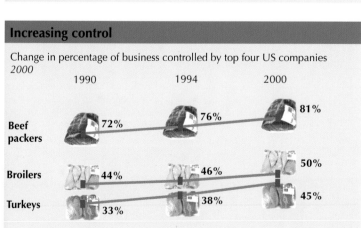

	1990	1994	2000
Beef packers	72%	76%	81%
Broilers	44%	46%	50%
Turkeys	33%	38%	45%

Pork production in USA

Number of sows produced by top four companies *2001*

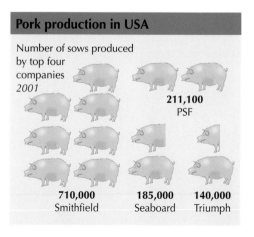

211,100
PSF

710,000
Smithfield

185,000
Seaboard

140,000
Triumph

Economies of scale

Supermarkets are getting ever larger – both in terms of turnover and floor area. The aim is to intensify sales – increase turnover per square yard/meter of sales space – in order to undercut smaller rivals.

THE POWER of the food retailers is immense. By a combination of sheer size, tight contracts and specifications, and the application of tough management techniques, farmers and even giant food processors now dance to the retailers' tune. Farmers' returns are being squeezed, and retailers, although on tight margins, are turning in huge profits.

In the UK, it has been estimated that half the food consumed by 57 million mouths is bought from just 1,000 stores. And in 1996, 35 percent of the £118 billion spent in the UK on food went as profit to the retail sector. In the USA, of every $1 spent by consumers in 2000, only 19 cents went to farmers and 81 cents went to everyone off the farm: truckers, processors, retailers, marketers.

Market economics has long been driven by a belief that maximum efficiency occurs when many producers vie for the attention of lots of informed consumers – a marketplace, bustling with traders and potential purchasers. But the 1980s saw the beginning of a new era of hypermarket economics and the food retail market is now dominated by a relatively small number of huge companies operating a strategy known as Efficient Consumer Response. ECR is the attempt to get every

element in the complex food-supply chain to operate in tandem, thereby maximizing efficiency. It is a triumph of the logistics industry, a revolution heralded by the arrival of EPOS: electronic point of sale technology – the ubiquitous laser bar-code scanning system. It is these that enable retailers to keep their back-up stocks to a minimum, reordering goods only as required, and thereby exerting immense power over the food chain. No longer do retailers carry the main burden of risk by having to buy goods up front and rely on them selling. Suppliers are now left waiting for orders that depend on the swipe of the bar-code scanner over their product at the supermarket checkout.

Some retailers have now taken even greater control, by reaching back down the food chain and contracting food processors to produce products that carry the retailer's own label. Some processors have prospered from such tie-ins, while others have been put out of business. In Europe, retailers' own-label products account for 45 percent of food sales. Outside Europe, by contrast, own-label products are typically less than 5 percent of total sales, with brand name products dominating.

Turnover
Average weekly sales per supermarket in USA
1973–2000
US$

$335,242
2000

$210,000
1995

$174,800
1990

$164,000
1985

$110,000
1980

$68,000
1975

Floor space
Average floor area of food retail store in USA
1990–2000
square feet

| 31,000 | 38,600 | 44,600 |
| 1990 | 1995 | 2000 |

Intensity of sales

Average turnover per square foot
1990–2000
US$

$391
2000

$293
1990

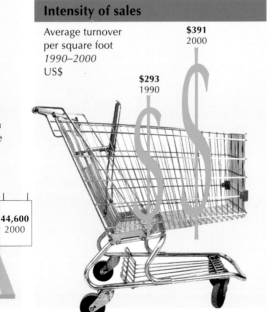

Top food retailers

Top retailers in USA

2001
US$

Kroger	$46.7bn
Safeway (US)	$31.5bn
Albertson's	$30.2bn
Wal-Mart	$28.2bn
Ahold	$24.1bn
Delhaize America	$13.2bn
Publix	$14.6bn
Winn-Dixie	$13.0bn

European retailers Go East

European retailers with outlets in Southeast and East Asia

- Carrefour (France)
- Makro (Netherlands)
- Tesco (UK)
- Metro (German)
- Auchan (France)

Until the 1990s retailers mostly catered to national needs, but by 2000 European retailers were opening stores overseas. The likely detrimental effect on smaller national retailers was recognized by the government in Malaysia when, in 2002, it attempted to prevent Tesco from building a series of hypermarkets.

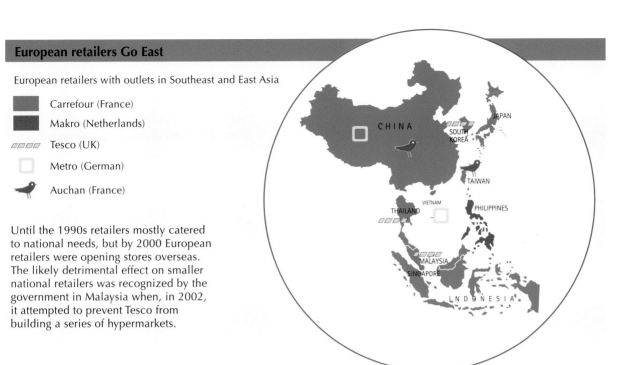

Wal-Mart takes over

Number of Wal-Mart owned stores and date of entry to country *2002*

The US-based Wal-Mart has 3,500 stores in the USA and a yearly turnover of $200 billion. It started its overseas expansion in 1991, and in 2002 planned to spend $9.2 billion on increasing its sales space worldwide.

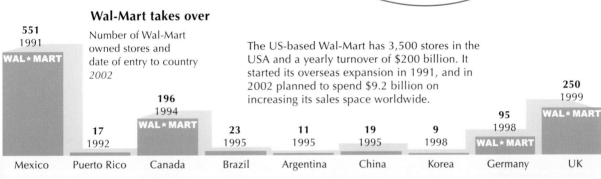

551 1991	**17** 1992	**196** 1994	**23** 1995	**11** 1995	**19** 1995	**9** 1998	**95** 1998	**250** 1999	
Mexico	Puerto Rico	Canada	Brazil	Argentina	China	Korea	Germany	UK	

Top retailers in UK

Share of food retail market
2001
Total market: £103 billion

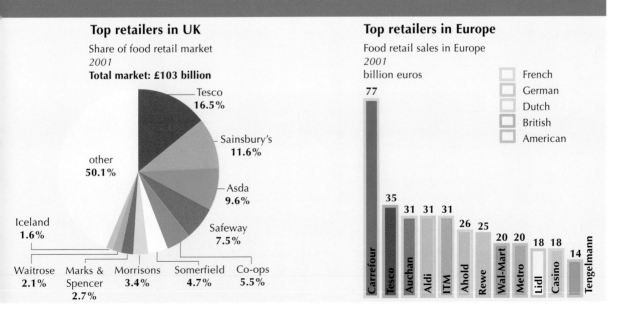

- Tesco **16.5%**
- Sainsbury's **11.6%**
- Asda **9.6%**
- Safeway **7.5%**
- Co-ops **5.5%**
- Somerfield **4.7%**
- Morrisons **3.4%**
- Marks & Spencer **2.7%**
- Waitrose **2.1%**
- Iceland **1.6%**
- other **50.1%**

Top retailers in Europe

Food retail sales in Europe
2001
billion euros

- French
- German
- Dutch
- British
- American

Carrefour	Tesco	Auchan	Aldi	ITM	Ahold	Rewe	Wal-Mart	Metro	Lidl	Casino	Tengelmann
77	35	31	31	31	26	25	20	20	18	18	14

35 | FUNCTIONAL FOODS

Expanding markets

The market for functional foods has expanded for several reasons:

- Scientific progress has led to better understanding of health benefits of some food.
- Lifestyle changes and expectations make some people choose what they buy and consume on the basis of the health benefits offered.
- The nutritional and health benefits of foodstuffs are seen by companies as an opportunity to achieve "added value", thereby increasing their profit margins in what are often static or declining traditional food and drink markets.

AN IMPORTANT GLOBAL TREND in food technology has been "functional foods" (also known as "nutraceuticals"). These are food or drink products that provide a health benefit beyond the nutrients traditionally contained in the food. They are promoted and sold on the basis of health-promoting benefits, often as the result of some added ingredient.

To the food industry functional foods assumed increasing importance from the mid-1990s onwards, and at the beginning of the 21st century the world's largest food and beverage companies are all either producing functional foods or trying to get into the market.

Some estimates put the potential global market for functional foods at around $50 billion. Others estimate that half of all food sold can be marketed as "functional", making the US market alone worth $250 billion. It partly depends on what is considered as a functional food/beverage.

Much market activity has been little more than "shoe-horning" ingredients (such as even more vitamins and minerals, but other ingredients as well) into existing or new products and claiming these will have an enhanced health benefit for consumers. There is often little detailed scientific substantiation behind these products as consumed that will guarantee that they will deliver long-term health benefits.

New insights from nutrition science have also enabled companies to stress the beneficial properties of traditional foods and drinks, repositioning these products in the market as "functional foods".

There are, however, some highly innovative functional food concepts that are carving out entirely new categories, based on real health benefits. Cardiovascular disease (CVD) is the single largest killer in the industrialized world, and since the discovery that natural plant sterols reduce LDL ("harmful") cholesterol levels, "heart health" products have been one of the most dynamic markets.

Other products have made a big impact on the market. These use strains of lactic acid bacteria (known as "probiotics"). These have scientifically documented health benefits when introduced into the gut's microflora.

Functional foods pose many complex challenges and marketing difficulties. Not least is the regulatory challenge. In the USA and European Union there is no legal definition for functional foods. Many functional foods and drinks fall into a gray regulatory area between food and drug. Most countries also have strict food-labeling laws that determine what can and what cannot be said about the health benefits of a particular food or ingredient.

Health-giving properties in natural foods

- **soluble fiber content of oats** – reduces LDL (harmful) cholesterol
- **lycopene in tomatoes** – possibly helps prevent some cancers
- **cranberry juice** – helps prevent urinary tract infections
- **soy protein** – promotes heart health
- **antioxidant vitamins** (A, C and E) – help prevent heart disease

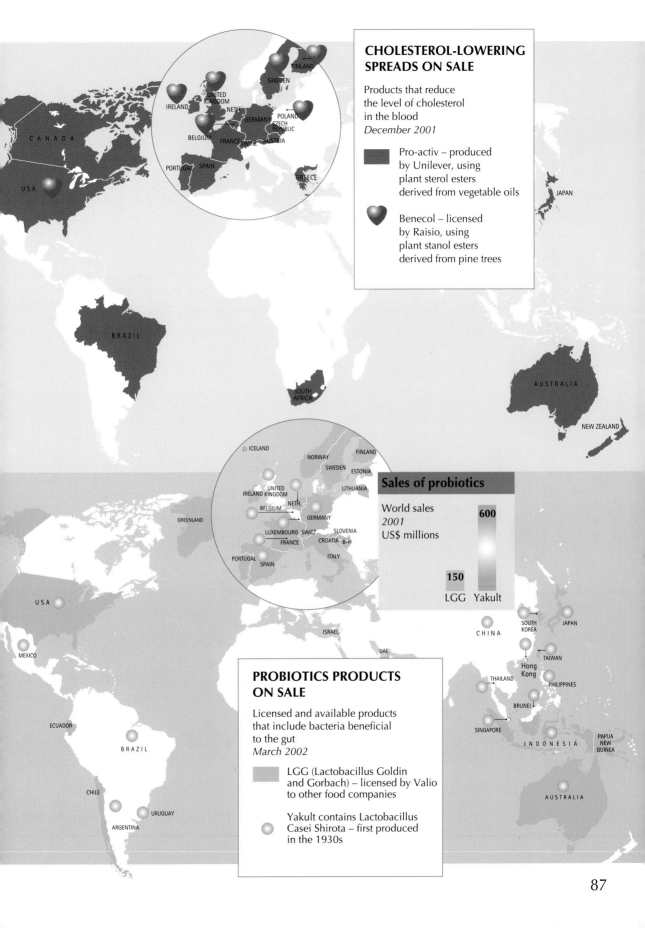

CHOLESTEROL-LOWERING SPREADS ON SALE

Products that reduce
the level of cholesterol
in the blood
December 2001

Pro-activ – produced
by Unilever, using
plant sterol esters
derived from vegetable oils

Benecol – licensed
by Raisio, using
plant stanol esters
derived from pine trees

Sales of probiotics

World sales
2001
US$ millions

600

150

LGG   Yakult

PROBIOTICS PRODUCTS ON SALE

Licensed and available products
that include bacteria beneficial
to the gut
March 2002

LGG (Lactobacillus Goldin
and Gorbach) – licensed by Valio
to other food companies

Yakult contains Lactobacillus
Casei Shirota – first produced
in the 1930s

87

Organic market

World sales
1997–2000
US$

$10bn

1997

$17.5bn

2000

CONSUMER DEMAND for organic produce in the industrialized world is growing steadily. There is an increased awareness of health and environmental issues, and higher disposable incomes are enabling people to make "lifestyle choices", such as paying more for food they feel will be better for them and less damaging to the environment. For example, public concerns about "mad cow disease" (BSE) in the UK and other countries in Europe (see page 36–37) have increased the demand for organic meat and milk.

The total world retail market for organic food and drink is estimated to have grown from $10 billion in 1997 to $17.5 billion in 2000. As yet it only represents between 1 percent and 3 percent of total retail sales, but a predicted annual growth rate of around 20 percent – as much as 25 percent in the UK – is causing excitement among companies eager to tap into this new and potentially lucrative market. Retailers are promoting organic food more vigorously, and the major food manufacturers are developing organic products to meet the growing demand.

The way in which organic produce is distributed varies between countries. Whereas in Germany most organic produce is sold through health-food shops and other specialist outlets, in the UK supermarkets have moved in and cornered 74 percent of the trade. In the USA many farmers sell their produce direct to the customer, and sales of organic produce in natural food stores are higher than sales in conventional ones.

Although the original ethos behind organic farming involved the purchase of locally produced food, in some countries much of the organic food now appearing in supermarkets has been imported. This not only includes tropical fruits and vegetables, and out-of-season produce, but in-season fruit and vegetables and grain, for which demand is too great to be met by national suppliers. Around 75 percent of the organic food consumed in the UK is imported, and in 2002 as much as 90 percent of organic food in Canada was supplied by the USA.

So while organic production saves energy on the farm (organic milk, for example, takes 20 percent of the energy input of ordinary milk), the transport of organic food for long distances (see pages 66 – 67) in most cases undoes many of the environmental benefits.

Consumer spending on organic food

Amount spent per person per year
2000
US$

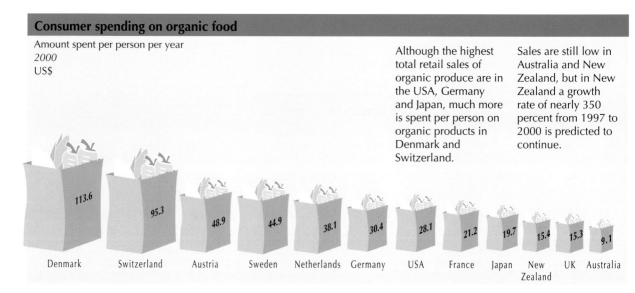

Denmark	Switzerland	Austria	Sweden	Netherlands	Germany	USA	France	Japan	New Zealand	UK	Australia
113.6	95.3	48.9	44.9	38.1	30.4	28.1	21.2	19.7	15.4	15.3	9.1

Although the highest total retail sales of organic produce are in the USA, Germany and Japan, much more is spent per person on organic products in Denmark and Switzerland.

Sales are still low in Australia and New Zealand, but in New Zealand a growth rate of nearly 350 percent from 1997 to 2000 is predicted to continue.

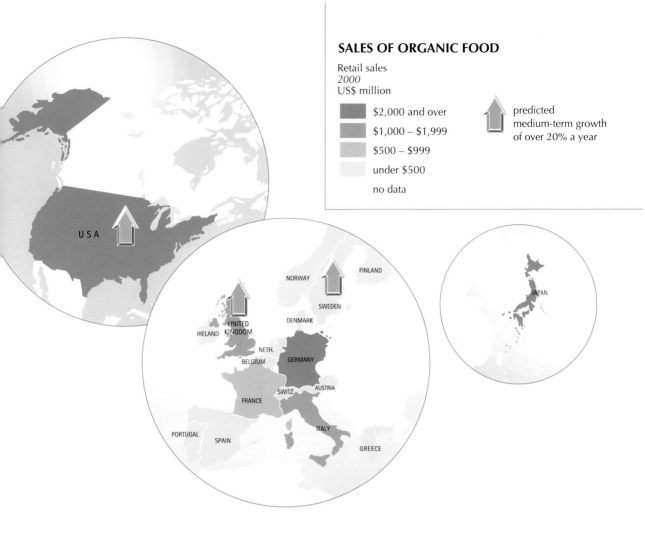

SALES OF ORGANIC FOOD

Retail sales
2000
US$ million

- $2,000 and over
- $1,000 – $1,999
- $500 – $999
- under $500
- no data

predicted medium-term growth of over 20% a year

USA

NORWAY
FINLAND
SWEDEN
DENMARK
UNITED KINGDOM
IRELAND
NETH.
BELGIUM
GERMANY
SWITZ.
AUSTRIA
FRANCE
ITALY
PORTUGAL
SPAIN
GREECE

JAPAN

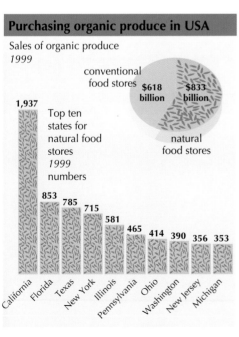

Purchasing organic produce in USA

Sales of organic produce
1999

conventional food stores **$618 billion**

$833 billion natural food stores

1,937 Top ten states for natural food stores *1999* numbers

State	Number
California	1,937
Florida	853
Texas	785
New York	715
Illinois	581
Pennsylvania	465
Ohio	414
Washington	390
New Jersey	356
Michigan	353

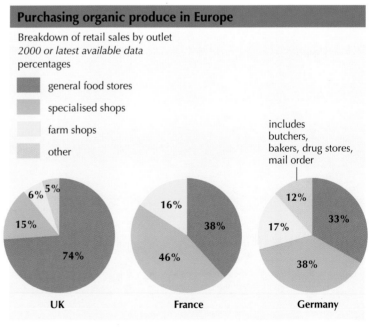

Purchasing organic produce in Europe

Breakdown of retail sales by outlet
2000 or latest available data
percentages

- general food stores
- specialised shops
- farm shops
- other

includes butchers, bakers, drug stores, mail order

UK
5%
6%
15%
74%

France
16%
38%
46%

Germany
12%
17%
33%
38%

37 | FOOD ADDITIVES

Human safety

540 food additive compounds are deemed safe for human consumption by regulatory bodies, but critics of the testing system have raised doubts about many of them:

320
accepted as reasonably safe

150
doubts and uncertainties raised about their safety

70
may cause allergy and / or acute intolerance in a few people

30
could cause significant long-term harm to any consumers

THE FOOD INDUSTRY spent around $20 billion in 2000 on chemical food additives to improve the color, flavor, texture and shelf life of its products. Consumers in the industrialized countries ingest between 13lb and 15lb (6kg – 7 kg) of food additives a year, for which the food industry is paying the equivalent of around $20 per person.

The food industry argues that its use of additives protects consumers from bacterial food poisoning and prevents foods from deteriorating. But the additives used as preservatives and those used to inhibit oils and fats from going rancid ("antioxidants") account for less than 1 percent by weight of the total quantity of additives in use. About 90 percent of all additives purchased by processing companies are "cosmetics" – they change the color, flavor, surface appearance, and texture of a food product. The remaining additives are "processing aids" such as lubricants and enzymes, used for their effect in the production process, rather then on the final product.

Flavorings are the single largest category of additives, with over 4,500 different compounds in use, compared with only 540 compounds in total used for other purposes. Artificial sweeteners are a large and distinct market, considered separate from flavorings.

The use of food additives is controlled by governments in all industrialized countries and in many developing countries too. Until

the 1950s this was done by "negative lists" – anything could be used unless it was specifically banned. Most countries have now developed "positive lists" – all chemicals not included on the list are banned. For a chemical to be included on the list, evidence must be provided that its risk to human health is acceptably slight.

As a rule, it is not acceptable to test food additives on humans, and so most tests are conducted using laboratory animals (usually rats and mice), and bacterial and cell cultures. The problem is that the results of animal studies are difficult to interpret because it is not known if animals or microbes provide suitable models for humans. The food additives industry often treats the results of animal studies as valid when they show no adverse effects, but questions their relevance when they do suggest adverse effects.

There are now positive lists for almost all categories of food additives except flavoring agents. The rules on the labeling of food additives vary considerably between countries, but one common practice is to require labeling of all additives drawn from positive lists, but not those from negative list systems. Consequently, additives such as colorings, preservatives and artificial sweeteners are often identified on labels, but the identity of flavoring compounds used in particular products is almost never disclosed.

Flavorings

Number of different compounds available
2002

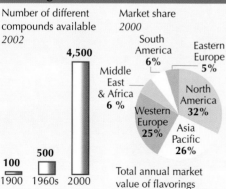

4,500

100 — 1900
500 — 1960s
2000

Market share
2000

South America 6%
Eastern Europe 5%
Middle East & Africa 6 %
North America 32%
Western Europe 25%
Asia Pacific 26%

Total annual market value of flavorings (excluding sweeteners)
$3.6 billion

Flavoring agents are used to reinforce the flavor of products containing natural foods or to simulate the taste of natural foods in products containing mainly starch and fat.

In most countries, flavorings (excluding sweeteners) are regulated less strictly than other kinds of food additives. They

do not have to be tested for safety and are only controlled or banned if shown to be harmful.

One reason given for this is that flavorings are used in small amounts (although this implies that they are powerful and reactive compounds). Another is that there are too many different ones to deal with each one

separately. Companies also argue for secrecy on the grounds of commercial confidentiality.

Savory and sweet food products

Savory and sweet foods are usually found on separate shelves in food stores, and are normally eaten as separate courses during a meal. But many processed products contain similar ingredients – fats, carbohydrates and starches – and their distinctive savory or sweet character is created by colorings and flavorings.

Colorings

Used to modify the color of the product. 40 compounds (or groups of compounds), of which 18 are synthetic.
The levels of colorings used in cereals, snack foods, dessert mixes, confectionery and candies and beverages are particularly high.

Preservatives

Chemicals added to foods to inhibit the growth of harmful micro-organisms. 50 compounds. Most common: sulfur dioxide, sulfites, sodium nitrate, sodium nitrite, potassium nitrate, potassium nitrite.

Antioxidants

Added to food products to inhibit oils and fats from going rancid.
17 compounds (or groups of compounds).

Emulsifiers and stabilizers

Used to ensure that water and oil remain mixed together. Used in margarine, mayonnaise, and also baked products.
48 compounds permitted in USA; 75 in EU.
Most common: lecithin, a natural ingredient of soybeans. As the cultivation of genetically modified soybeans increases, so too does the proportion of lecithin obtained from GM-sources. Stabilizers are added to emulsions to hold them together and to prevent them for separating out.

Thickeners

Used to thicken a wide range of products, including emulsions. Most widely used: starch. Starches are modified to control their performance in different kinds of products. There are at least 40 different chemical methods for modifying starches.

Anti-caking agents

Used to inhibit absorption of water and to prevent powdered mixtures sticking together.

Flavor enhancers

Chemicals that trick the taste buds into thinking a food has more flavor than it does. 36 compounds. Best-known: MSG (monosodium glutamate), used in savory products. Approximately 650,000 tonnes of MSG are added to the world's food supply each year.

Sweeteners

Excluding sugar there are 13 compounds or groups of compounds that are used as artificial sweeteners. Synthetic sweeteners, which include acesulfame-K, aspartame, saccharin and sucralose, provide virtually no calories. Approximately 15,000 tonnes are used each year worldwide. Nutritive sweeteners, such as lactose and glucose syrup, are used in even greater quantities.

The global market for artificial sweeteners was estimated in 2000 as $2.5 billion. The market leader is aspartame, of which more than $1 billion is sold annually.

Use of aspartame
1982–1995
tonnes

13,140
10,000
5,730
370

1982 1986 1991 1995

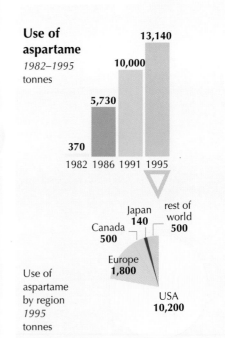

Use of aspartame by region
1995
tonnes

Japan 140
Canada 500
rest of world 500
Europe 1,800
USA 10,200

38 EATING OUT

Increased spending

Total spent each year on eating out
1980–2000
US$ billion

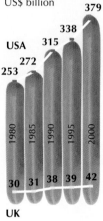

USA
253 272 315 338 379

1980 1985 1990 1995 2000

UK
30 31 38 39 42

$1,400

is the average
**spent each year
per person**
in the USA
on food
eaten out
of the home

EATING OUT is a trillion-dollar global business and the USA is a key player. Each year, Americans spend almost $1,400 per person on eating in restaurants, fast food outlets, hotels, schools and at work.

In most developed countries, annual expenditure on eating out exceeds $400 per person. International trends are clearly apparent in the menus that are on offer, with pizza, burgers and fried chicken becoming widely available in developed and developing countries alike. "Fast food", with its casual, informal, and relatively quick service, is also becoming available worldwide. However, despite the global reach of the fast-food chains, eating out is a long-established activity in all countries, whether rich or poor, and local culture still influences the style of food on offer in each country. In Italy, for example, the tradition of family-run restaurants means that it is difficult for international players to gain a significant foothold.

Eating out is generally a more expensive option than eating at home, so it tends to be more viable, and therefore more prevalent, in countries with higher levels of disposable income. But there are exceptions at both ends of the economic scale. In the most advanced economies there are many opportunities for eating out at low cost. At the other economic extreme, where many homes may be without a kitchen, there is considerable demand for low-cost cooked food readily available in the street. In Southeast and East Asia, the tradition of buying snacks from street traders continues to influence the market: the expenditure on, and frequency of, eating out is especially high in Japan, for example.

CANADA

USA

MEXICO

VENEZUELA

COLOMBIA

BRAZIL

CHILE

URUGUAY

ARGENTINA

Restaurant size

Average turnover of food-service outlets
US$ thousands
2000

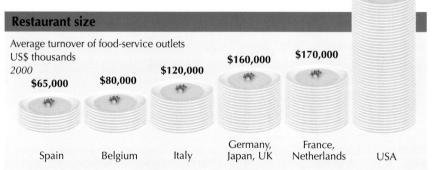

$65,000 — Spain
$80,000 — Belgium
$120,000 — Italy
$160,000 — Germany, Japan, UK
$170,000 — France, Netherlands
$450,000 — USA

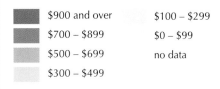

MONEY SPENT ON MEALS OUT

Expenditure on food consumed
away from home per person per year
US$ *2000*

■ $900 and over		$100 – $299
■ $700 – $899		$0 – $99
■ $500 – $699		no data
■ $300 – $499		

Number of meals eaten out

per person per year

Japan	USA	UK
198	197	144

Germany	France	Belgium
139	133	129

Spain	Italy	Netherlands
124	120	115

Eating out is a growing market. Expenditure in the USA increased by 50 percent between 1980 and 2000, and by 40 percent in the UK. In 2000, half of the amount spent on food in the USA was spent on eating away from home. Other industrialized countries – including France, Germany and the UK – spent less than a third of their food budgets on eating out, but the signs are that they will increase to US levels by 2020 or 2025. The rate of increase of eating out in the USA and UK is influenced by factors such as the number of single-parent households and the ratio of younger to older people in the population. It is also directly paralleled by growth in the number of women in paid work.

Although restaurants in the USA have learnt to maximize turnover, average expenditure per person and turnover per outlet vary considerably within Europe, largely for cultural reasons. Many German diners enjoy large, bustling restaurants, whereas in Spain, Italy, and Belgium, smaller, family-run businesses are more common.

Microwave sales

Number of Sharp microwaves sold
1962–2002
cumulative total

0
1962

5 million
1981

70 million

2000

Ready rise

Increase in UK spending on ready meals
2001–2002

8%

frozen

24%

chilled

FAST FOOD is a part of all cultures. A meat burger in a bun, a frankfurter sausage, fish and chips, a pizza slice, samosas, deep-fried dumplings, water melon can all be eaten quickly, out of the home, often with the hands. But it has been the global spread of the three big US fast-food corporations – McDonalds, Burger King and Tricon (Kentucky Fried Chicken and Pizza Hut) – that has come to represent fast food and influenced our eating habits.

Fast food is central to the American way of life. Each day one in five Americans eats in a fast-food restaurant. Elsewhere in the world the figures are lower, but everywhere the trend is upwards, with Latin America and Asia experiencing the biggest increase in McDonald's restaurants in the late 1990s. National companies may successfully imitate US style, as with Jollibee in the Philippines, which had an annual turnover in 2000 of US$390 million through 406 outlets, and has now expanded into the USA.

Not all "fast food" is eaten out of the home, however. The microwave was discovered in 1946, and by 1975 more microwave ovens were being sold than gas cookers worldwide. They took off in Japan and helped spawn a global market in chilled or frozen "ready meals". Providing almost instant gratification, they have become deeply embedded in the food culture of the industrialized world, where the family meal time has been replaced by "grazing" – eating when it is convenient, usually in front of the TV.

Fast food and ready meals tend to be high in animal fats. Their popularity has contributed to the rising levels of obesity in the industrialized world – in particular in the USA, where 44 million people are now considered obese, and a further 6 million super-obese (more than 100 pounds/45 kg overweight). Not all fast food is unhealthy, however. In the wake of scares over the safety of meat in Europe manufacturers are beginning to respond to consumers' demands for meat-free ready meals. The demand for these products in the UK rose by 16 percent in 2001.

McDonald's

Regional increase in number of restaurants
1996–2001

8%
USA

23%
Canada

76%
Europe

126%
Asia, Pacific, Middle East and Africa

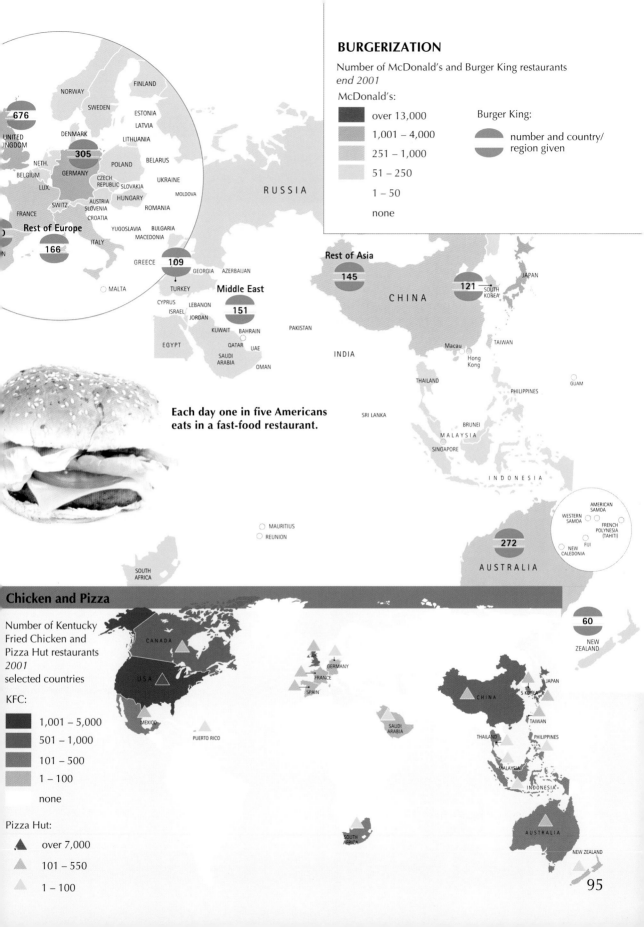

BURGERIZATION

Number of McDonald's and Burger King restaurants
end 2001

McDonald's:

- over 13,000
- 1,001 – 4,000
- 251 – 1,000
- 51 – 250
- 1 – 50
- none

Burger King:
- number and country/region given

676
UNITED KINGDOM

NORWAY
SWEDEN
FINLAND
ESTONIA
LATVIA
LITHUANIA

DENMARK

305
GERMANY

NETH.
BELGIUM
LUX.
POLAND
BELARUS
CZECH REPUBLIC
SLOVAKIA
UKRAINE

FRANCE
SWITZ.
AUSTRIA
SLOVENIA
CROATIA
HUNGARY
MOLDOVA
ROMANIA

Rest of Europe

166
ITALY
YUGOSLAVIA
MACEDONIA
BULGARIA

RUSSIA

MALTA

GREECE **109**

GEORGIA
AZERBAIJAN

TURKEY

CYPRUS
LEBANON
ISRAEL
JORDAN

Middle East

151

KUWAIT
BAHRAIN
EGYPT
QATAR
UAE
SAUDI ARABIA
OMAN

PAKISTAN

Rest of Asia

145

CHINA

121 SOUTH KOREA

JAPAN

INDIA

Macau
Hong Kong

TAIWAN

THAILAND

GUAM

PHILIPPINES

SRI LANKA

BRUNEI
MALAYSIA
SINGAPORE

INDONESIA

Each day one in five Americans eats in a fast-food restaurant.

MAURITIUS
REUNION

SOUTH AFRICA

AMERICAN SAMOA
WESTERN SAMOA
FRENCH POLYNESIA (TAHITI)
FIJI
NEW CALEDONIA

272
AUSTRALIA

60
NEW ZEALAND

Chicken and Pizza

Number of Kentucky Fried Chicken and Pizza Hut restaurants
2001
selected countries

KFC:

- 1,001 – 5,000
- 501 – 1,000
- 101 – 500
- 1 – 100
- none

Pizza Hut:

- ▲ over 7,000
- ▲ 101 – 550
- ▲ 1 – 100

CANADA
USA
MEXICO
PUERTO RICO

UK
GERMANY
FRANCE
SPAIN

SAUDI ARABIA

CHINA
JAPAN
S KOREA
TAIWAN
THAILAND
PHILIPPINES
MALAYSIA
INDONESIA

SOUTH AFRICA

AUSTRALIA
NEW ZEALAND

Regional consumption

Total alcohol consumption per person per year
2000
liters of pure alcohol

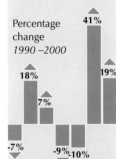

8	7	4	7	8	2	4
Western Europe	Eastern Europe	Latin America	North America	Australasia	rest of world	total

Percentage change
1990–2000

18%
7%
-7%
-9% -10%
41%
19%

WORLDWIDE, CONSUMPTION of alcoholic drink averages about four liters of pure alcohol per person per year. Although this has been relatively constant over the last 20 years, levels are rising in many developing countries.

Around 140 million people suffer from alcohol dependence, including 14 million Americans. It is estimated that in 1990 alcohol was responsible for 3.5 percent of the global burden of disease and disability, although this varies considerably between regions. In India, and in Islamic countries where alcohol is prohibited, the problem is negligible. In Europe, one in four deaths of men aged between 15 and 29 years is related to alcohol, and one in three in parts of Eastern Europe.

Figures for alcohol consumption vary in reliability and tell only part of the story. Of equal importance is who drinks, and how they drink. In wine-drinking Italy and Spain, heavy drinkers tend to spread their drinking evenly, reducing acute problems although they are still susceptible to long-term consequences. In the UK and Ireland drinkers are more likely to binge drink, increasing the risk of acute problems. They also start "problem" drinking younger. Countries such as South Africa,

where average alcohol consumption appears low, may still have a problem if the minority who drink do so at very high levels.

The combined sales of the top 10 global brewers and top 10 distilled spirits companies total nearly $200 billion. Because they contribute large amounts to governments in alcohol taxes, they are a powerful political force in many countries. There are concerns about their activities in India, where local controls are giving way to a deregulated market. In Europe, advertisements for energy drinks and premixed "alcopops" successfully link alcohol with youth lifestyle and sexual attraction, thereby encouraging drinking to become established as a habit early on in life.

Market controls can be effective. OECD countries that ban alcohol advertising record consumption about 16 percent lower, and traffic fatalities 23 percent lower, than those with no restrictions. Most countries operate age limits on sales of alcohol ranging from 16 years of age in Belgium, France, Italy and Spain, to 21 years in Korea, Malaysia, Ukraine and some US states, but others, including Azerbaijan, China, Georgia, Portugal and Thailand, have no established legal limits.

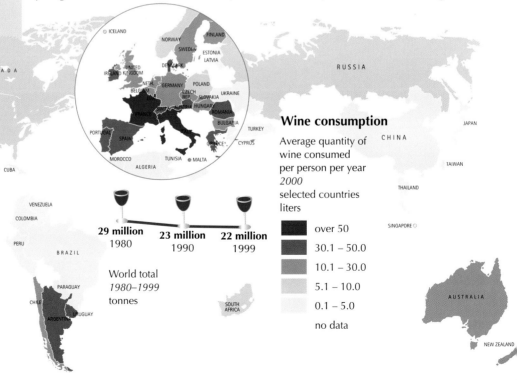

29 million 1980 **23 million** 1990 **22 million** 1999

World total
1980–1999
tonnes

Wine consumption

Average quantity of wine consumed per person per year
2000
selected countries
liters

over 50	
30.1 – 50.0	
10.1 – 30.0	
5.1 – 10.0	
0.1 – 5.0	
no data	

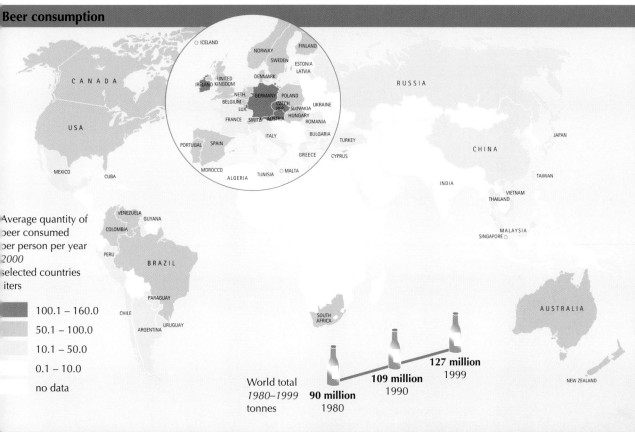

Beer consumption

Average quantity of beer consumed per person per year
2000
selected countries
liters

- 100.1 – 160.0
- 50.1 – 100.0
- 10.1 – 50.0
- 0.1 – 10.0
- no data

World total
1980–1999
tonnes

90 million
1980

109 million
1990

127 million
1999

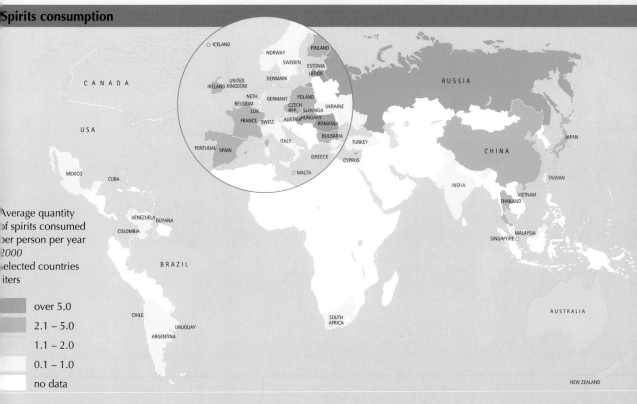

Spirits consumption

Average quantity of spirits consumed per person per year
2000
selected countries
liters

- over 5.0
- 2.1 – 5.0
- 1.1 – 2.0
- 0.1 – 1.0
- no data

41 | ADVERTISING

Big spenders

Amount spent
in 1999
by world's biggest
food advertisers
2000
US$ billion

Kellogg's	0.6
Danone	0.7
Pepsi	1.1
Mars	1.2
McDonald's	1.5
Coca-Cola	1.9
Nestlé	

PEOPLE DO NOT NEED to be persuaded to eat, and yet the food industry spends $40 billion each year advertising its products.

It is not usually fresh foods that are advertised, but branded and processed products. These are higher in sugars, refined starches, fats, and added salt compared with fresh unprocessed foods, and are seen by health practitioners to encourage unhealthy diets. But for every dollar spent by the World Health Organization on trying to improve the nutrition of the world's population, around $500 is spent by the food industry on promoting processed foods.

Coca-Cola is the most heavily advertised brand, with an annual advertising spend of $1.5 billion worldwide. As with fizzy drinks, processed foods are marketed to people who can afford to buy them – that is, wealthier countries with higher incomes. During the 1990s, new markets opened up with the emerging economies of Eastern Europe, and companies such as Procter & Gamble, Nestlé, Mars, and Unilever were quick to re-focus their attentions and increase promotion of their branded products. In Russia, the ten largest advertisers spent $239 million in 1999 – an increase of 30 percent on 1995 spending.

The food industry spends

$40 billion

each year on advertising

Money spent on food advertising

Countries' shares
of annual world total
1998 and 1999

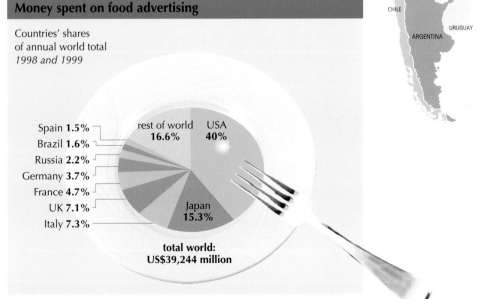

Spain **1.5%**
Brazil **1.6%**
Russia **2.2%**
Germany **3.7%**
France **4.7%**
UK **7.1%**
Italy **7.3%**

rest of world **16.6%**
USA **40%**
Japan **15.3%**

total world:
US$39,244 million

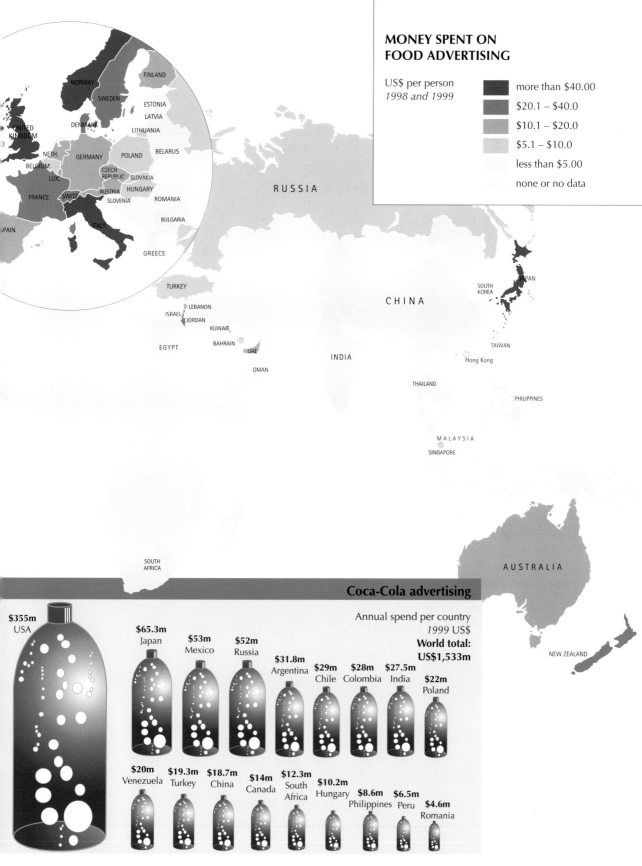

MONEY SPENT ON FOOD ADVERTISING

US$ per person
1998 and 1999

■	more than $40.00
■	$20.1 – $40.0
■	$10.1 – $20.0
■	$5.1 – $10.0
	less than $5.00
	none or no data

Coca-Cola advertising

Annual spend per country
1999 US$
**World total:
US$1,533m**

$355m
USA

$65.3m
Japan

$53m
Mexico

$52m
Russia

$31.8m
Argentina

$29m
Chile

$28m
Colombia

$27.5m
India

$22m
Poland

$20m
Venezuela

$19.3m
Turkey

$18.7m
China

$14m
Canada

$12.3m
South Africa

$10.2m
Hungary

$8.6m
Philippines

$6.5m
Peru

$4.6m
Romania

FOOD AND WATER are vital ingredients for life. It is little wonder, then, that access to safe, nutritious food has become a political issue worldwide.

The production, distribution and retailing of food is, in many countries, controlled by large corporations and governments. Yet, consumers of that food – and the people most closely involved in producing, transporting, serving, selling and inspecting it – have also found a voice, and have formed large umbrella groups that link affiliated organizations around the world.

Many food issues are the focus of sustained campaigning. They include access to safe food and water, the problem of contaminated food and water, the impact on health of a changing diet, animal welfare, the impact on the environment of unsustainable farming practices, worker health and safety, wages and conditions, and the techniques employed to advertise food.

International Union (IUF)

Number of unions affiliated to the International Union of Food, Agricultural, Hotel, Restaurant, Catering, Tobacco and Allied Workers' Associations (IUF) *2002*

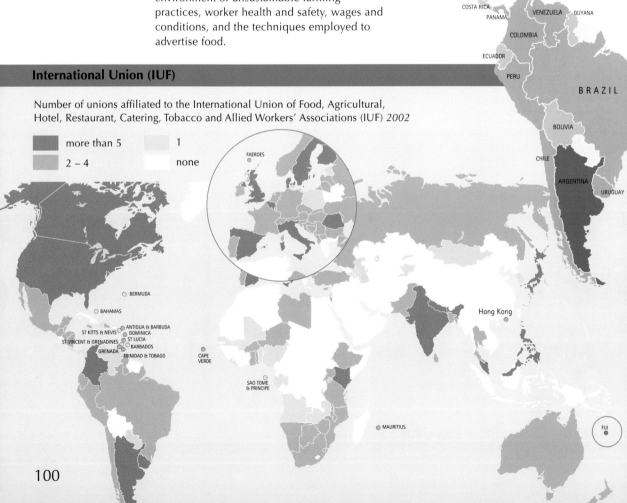

- more than 5
- 2 – 4
- 1
- none

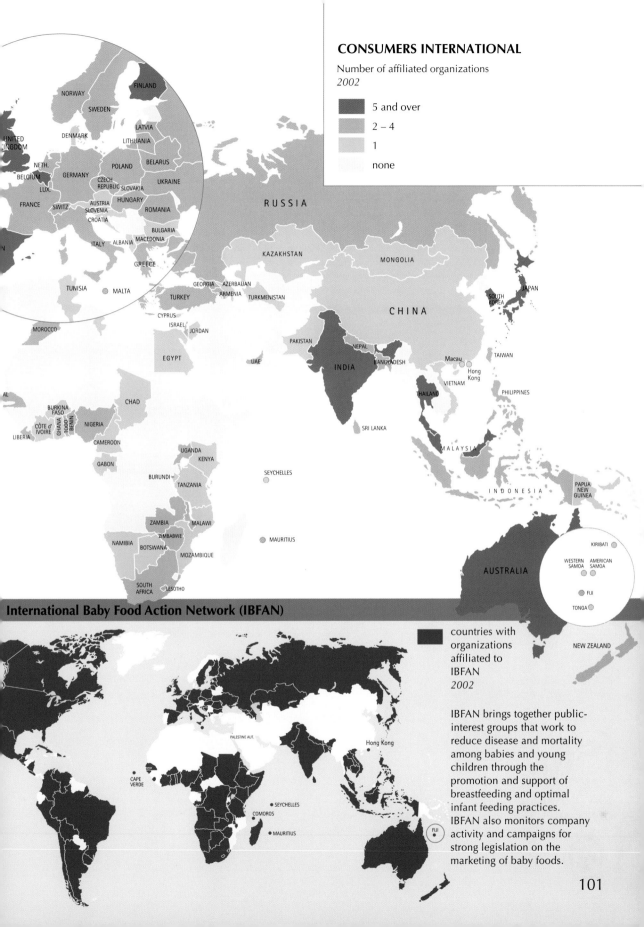

CONSUMERS INTERNATIONAL

Number of affiliated organizations
2002

- 5 and over
- 2 – 4
- 1
- none

RUSSIA

NORWAY
SWEDEN
FINLAND
LATVIA
LITHUANIA
DENMARK
UNITED KINGDOM
NETH.
BELGIUM
LUX.
GERMANY
POLAND
BELARUS
UKRAINE
CZECH REPUBLIC
SLOVAKIA
FRANCE
SWITZ.
AUSTRIA
SLOVENIA
HUNGARY
ROMANIA
CROATIA
BULGARIA
ITALY
ALBANIA
MACEDONIA
GREECE
TUNISIA
MALTA
MOROCCO
GEORGIA
AZERBAIJAN
TURKEY
ARMENIA
TURKMENISTAN
CYPRUS
ISRAEL
JORDAN
EGYPT
UAE

KAZAKHSTAN
MONGOLIA
CHINA

SOUTH KOREA
JAPAN
TAIWAN
PAKISTAN
NEPAL
BANGLADESH
INDIA
Macau
Hong Kong
VIETNAM
THAILAND
SRI LANKA
PHILIPPINES
MALAYSIA
INDONESIA

CHAD
BURKINA FASO
CÔTE d' IVOIRE
GHANA
TOGO
BENIN
NIGERIA
LIBERIA
CAMEROON
GABON
UGANDA
KENYA
BURUNDI
TANZANIA
SEYCHELLES
ZAMBIA
MALAWI
ZIMBABWE
NAMIBIA
BOTSWANA
MOZAMBIQUE
MAURITIUS
SOUTH AFRICA
LESOTHO

PAPUA NEW GUINEA
AUSTRALIA
KIRIBATI
WESTERN SAMOA
AMERICAN SAMOA
FIJI
TONGA
NEW ZEALAND

International Baby Food Action Network (IBFAN)

- countries with organizations affiliated to IBFAN
2002

PALESTINE AUT.
Hong Kong
CAPE VERDE
SEYCHELLES
COMOROS
MAURITIUS
FIJI

IBFAN brings together public-interest groups that work to reduce disease and mortality among babies and young children through the promotion and support of breastfeeding and optimal infant feeding practices. IBFAN also monitors company activity and campaigns for strong legislation on the marketing of baby foods.

WORLD TABLES

"It is not the role of the manufacturing industry to improve the health of the general public or to shoulder the responsibility of ensuring that people live longer, or lead healthier lives."
– Tim Fortesque, Secretary General of the Food & Drinks Industries Council, UK, quoted by M R Turner in 1981

AGRICULTURE

Countries	1 LAND AREA *1998* 1,000 hectares	2 CROPLAND			3 AGRICULTURAL WORKERS As percentage of total workforce *2000*	4 TRACTORS Per 1,000 people *1998*
		AREA *1997* 1,000 hectares	IRRIGATED LAND Percentage of cropland *1997*	CROP YIELD kg per hectare *1996–98*		
Afghanistan	65,209	8,054	35%	1,363	67%	0.1
Albania	2,740	702	48%	2,584	48%	10.8
Algeria	238,174	8,040	7%	980	24%	37.1
Angola	124,670	3,500	2%	642	72%	2.5
Argentina	273,669	27,200	6%	3,260	10%	34.1
Armenia	2,820	559	52%	1,772	13%	74.8
Australia	768,230	53,100	5%	1,995	5%	711.1
Austria	8,273	1,479	0%	5,659	5%	1,735.8
Azerbaijan	8,660	1,935	75%	1,647	27%	35.3
Bahamas	1,390	–	0%	–	4%	18.3
Bahrain	68	–	0%	–	1%	–
Bangladesh	13,017	8,241	45%	2,705	56%	0.1
Belarus	20,748	6,319	2%	2,204	13%	128.7
Belgium	3,023	785	4%	–	2%	1,262.7
Belize	2,280	89	3%	2,096	30%	46.0
Benin	11,062	1,595	1%	1,096	54%	0.1
Bhutan	4,700	160	25%	1,097	94%	–
Bolivia	108,438	2,100	4%	1,592	44%	0.1
Bosnia-Herzegovina	5,100	650	0%	2,382	5%	290.0
Botswana	56,673	346	0%	258	45%	19.1
Brazil	845,651	65,300	5%	2,458	17%	4.0
Brunei	580	–	0%	–	1%	72.0
Bulgaria	11,055	4,511	18%	2,644	7%	77.9
Burkina Faso	27,360	3,440	1%	729	92%	0.4
Burma	65,755	10,151	15%	2,947	70%	0.5
Burundi	2,568	1,100	1%	1,337	90%	0.1
Cambodia	17,652	3,807	7%	1,785	70%	0.3
Cameroon	46,540	7,160	0%	1,274	59%	0.1
Canada	922,097	45,700	2%	2,737	2%	1,756.4
Central African Rep.	62,298	2,020	0%	1,002	73%	0.1
Chad	125,920	3,256	1%	647	75%	0.1
Chile	74,880	2,297	55%	4,455	16%	9.1
China	932,641	135,365	38%	4,837	68%	1.4
Colombia	103,870	4,430	24%	2,798	20%	0.8
Comoros	220	–	0%	–	74%	–
Congo	34,150	185	1%	707	41%	1.4
Congo, Dem. Rep.	226,705	7,880	0%	746	63%	0.2
Costa Rica	5,106	505	25%	3,162	20%	21.4
Côte d'Ivoire	31,800	7,350	1%	1,102	49%	1.3
Croatia	5,592	1,442	0%	4,733	8%	13.8
Cuba	10,982	4,450	20%	1,973	14%	98.2
Cyprus	930	–	0%	–	9%	500.0

 Sources: Col 1: *World Resources 1998–1999;* **Col 2:** *World Resources 2000–2001, Tables AF.1 and AF.2;* **Col 3:** FAO, database <http://apps.fao.org/>;
Col 4: FAO, *Production Yearbook 2000*

5 AGROCHEMICALS		6 LIVESTOCK PRODUCTION				7 FISH	Countries
PESTICIDES kg used per hectare of cropland 1996	FERTILIZERS kg used per hectare of cropland 1995–97	CHICKENS	PIGS	BEEF AND VEAL	LAMB AND MUTTON	Total production 1999	
–	1	0.75	–	32.16	275.67	1,200	Afghanistan
435	7	2.57	33.60	108.49	433.95	3,055	Albania
835	8	6.36	0.10	21.56	318.58	105,943	Algeria
42	1	0.65	34.17	44.92	6.47	177,497	Angola
1,266	28	18.59	72.91	332.28	119.33	1,026,022	Argentina
2	14	1.42	38.55	69.21	86.61	877	Armenia
2,535	43	20.75	266.11	451.57	1,745.95	250,075	Australia
2,710	168	8.81	666.96	83.10	37.13	3,502	Austria
–	14	1.65	2.46	60.65	253.58	4,935	Azerbaijan
–	–	26.06	11.40	0.26	4.93	10,474	Bahamas
–	–	5.83	–	11.67	471.88	10,295	Bahrain
176	138	1.09	–	17.86	2.76	1,544,170	Bangladesh
–	157	5.03	292.40	153.63	10.21	5,809	Belarus
–	1,475	23.60	1,038.52	78.70	20.40	30,722	Belgium
17,804	50	19.50	141.08	26.52	2.74	43,103	Belize
–	22	4.44	34.97	31.09	38.46	38,542	Benin
670	1	0.22	15.68	32.47	10.79	330	Bhutan
1,514	5	8.60	183.51	111.88	230.42	6,450	Bolivia
–	11	0.41	41.04	18.61	56.58	2,500	Bosnia-Herzegovina
40	9	5.18	9.25	129.79	74.63	2,000	Botswana
836	78	25.69	133.73	182.76	26.23	774,750	Brazil
–	–	20.73	13.72	96.65	–	3,308	Brunei
966	43	8.15	364.74	46.55	283.05	18,336	Bulgaria
1	10		30.16	40.75	129.17	7,660	Burkina Faso
16	19	4.31	41.43	17.80	2.74	945,827	Burma
268	3	0.90	12.25	10.40	13.22	9,254	Burundi
–	2	1.75	188.04	36.06	–	284,100	Cambodia
253	5	1.99	39.77	43.49	90.75	95,067	Cameroon
644	67	19.58	640.93	128.10	17.10	1,135,516	Canada
12	0	1.11	103.73	98.20	22.87	15,117	Central African Rep.
223	3	0.87	2.09	77.93	91.31	84,000	Chad
3,240	210	18.08	213.66	61.82	51.73	5,324,744	Chile
–	265	4.95	432.94	28.18	78.22	40,029,919	China
6,134	125	7.92	52.93	90.27	11.35	170,896	Colombia
–	–	0.90	–	12.75	7.79	12,200	Comoros
216	20	1.97	12.23	4.64	10.93	43,886	Congo
–	1	5.00	15.87	1.70	5.63	208,862	Congo, Dem. Rep.
18,726	322	11.19	87.99	93.69	0.19	35,003	Costa Rica
–	16	3.25	17.04	18.19	34.35	77,000	Côte d'Ivoire
3,060	161	5.72	196.74	31.37	29.87	25,519	Croatia
–	52	4.91	160.70	43.90	8.22	122,425	Cuba
–	–	23.28	801.53	23.34	207.91	6,695	Cyprus

5: *World Resources 2000–2001*, Table AF.2; **Col 6:** FAO, database <http://apps.fao.org/>; **Col 7:** FAO *World fisheries production by capture and culture by country*, 1999 <ftp://ftp.fao.org/fi/stat/summ_99/aqua_a-0.pdf>

AGRICULTURE

Countries	1 LAND AREA *1998* 1,000 hectares	2 CROPLAND			3 AGRICULTURAL WORKERS As percentage of total workforce *2000*	4 TRACTORS Per 1,000 people *1998*
		AREA *1997* 1,000 hectares	**IRRIGATED LAND** Percentage of cropland *1997*	**CROP YIELD** kg per hectare *1996–98*		
Czech Republic	7,728	3,331	1%	4,178	8%	175.5
Denmark	4,243	2,373	20%	6,130	4%	1,217.4
Djibouti	2,320	–	0%	–	–	0.1
Dominican Republic	4,838	1,500	17%	3,851	17%	3.7
Ecuador	27,684	3,001	8%	1,746	26%	7.0
Egypt	99,545	3,300	100%	6,681	33%	10.4
El Salvador	2,072	816	15%	1,906	29%	4.4
Equatorial Guinea	2,805	230	0%	–	70%	0.8
Eritrea	10,100	393	7%	451	78%	0.4
Estonia	4,227	1,143	0%	1,978	11%	559.4
Ethiopia	100,000	10,500	2%	1,206	82%	0.1
Fiji	1,827	285	1%	2,094	40%	54.3
Finland	30,459	2,129	3%	3,381	5%	1,298.3
France	55,010	19,468	9%	7,109	3%	1,355.4
Gabon	25,767	495	1%	1,742	38%	7.0
Gambia	1,000	200	1%	1,006	79%	0.1
Georgia	6,970	1,066	44%	–	20%	32.9
Germany	34,927	12,060	4%	6,366	3%	1,005.8
Ghana	22,754	4,550	0%	1,365	57%	0.7
Greece	12,890	3,915	35%	3,535	17%	3,003.5
Guatemala	10,843	1,905	7%	1,882	46%	2.3
Guinea	24,572	1,485	6%	1,297	84%	0.2
Guinea-Bissau	2,812	350	5%	1,410	83%	0.0
Guyana	19,685	496	26%	3,982	18%	53.4
Haiti	2,756	910	10%	969	62%	0.1
Honduras	11,189	2,045	4%	1,528	32%	6.7
Hungary	9,234	5,047	4%	4,275	11%	173.4
Iceland	10,025	6	0%	–	8%	809.2
India	297,319	169,850	34%	2,204	60%	5.9
Indonesia	181,157	30,987	16%	3,915	48%	1.4
Iran	162,200	19,400	37%	1,863	26%	37.0
Iraq	43,737	5,540	64%	835	10%	76.0
Ireland	6,889	1,346	0%	6,721	10%	1,021.3
Israel	2,062	437	46%	1,776	3%	331.1
Italy	29,406	10,927	25%	4,920	5%	1,044.6
Jamaica	1,083	274	12%	1,227	21%	11.3
Japan	37,652	4,295	63%	6,023	4%	758.1
Jordan	8,893	390	19%	1,037	11%	28.2
Kazakhstan	267,073	30,135	7%	670	18%	44.8
Kenya	56,914	4,520	1%	1,535	75%	1.2
Kirgistan	19,180	1,425	75%	–	26%	36.0
Korea (North)	12,041	2,000	73%	2,251	30%	19.7

Sources: Col 1: *World Resources 1998–1999;* **Col 2:** *World Resources 2000–2001, Tables AF.1 and AF.2;* **Col 3:** FAO, database <http://apps.fao.org/>; **Col 4:** FAO, *Production Yearbook 2000*

5 AGROCHEMICALS		6 LIVESTOCK PRODUCTION				7 FISH	Countries
PESTICIDES kg used per hectare of cropland 1996	FERTILIZERS kg used per hectare of cropland 1995–97	CHICKENS	PIGS	BEEF AND VEAL	LAMB AND MUTTON	Total production 1999	
1,169	107	15.33	425.21	36.45	8.42	22,965	Czech Republic
2,200	167	26.58	3,986.40	116.95	12.88	1,447,664	Denmark
–	–	–	–	52.22	324.37	350	Djibouti
–	59	24.72	110.65	41.80	4.24	9,269	Dominican Republic
1,696	48	10.75	177.47	79.27	35.43	625,247	Ecuador
1,293	343	6.52	1.04	20.33	43.46	606,780	Egypt
2,642	106	5.58	25.07	33.07	0.33	15,467	El Salvador
–	0	0.47	6.07	0.90	18.38	7,001	Equatorial Guinea
–	11	0.47	–	40.99	155.78	7,042	Eritrea
105	29	8.60	308.02	76.81	10.70	111,993	Estonia
34	16	1.48	0.44	43.71	131.94	15,858	Ethiopia
2,333	66	7.59	97.92	56.68	2.58	38,471	Fiji
410	147	9.66	414.66	69.59	6.63	176,018	Finland
–	262	14.13	450.24	91.48	124.51	845,649	France
–	1	3.75	89.72	6.75	46.75	53,440	Gabon
46	5	0.88	6.90	22.19	20.46	30,004	Gambia
–	33	1.77	81.52	51.69	100.72	1,600	Georgia
2,085	250	4.86	538.80	52.25	26.38	312,492	Germany
2,333	4	0.90	14.10	10.12	32.68	493,206	Ghana
–	133	12.96	208.74	28.05	680.22	215,964	Greece
574	99	6.68	18.88	30.74	14.93	15,878	Guatemala
274	3	0.45	4.71	26.24	26.24	87,314	Guinea
83	1	1.15	219.29	34.20	60.05	5,000	Guinea-Bissau
–	29	13.01	11.61	26.28	68.33	54,450	Guyana
23	9	0.97	54.73	32.42	5.16	5,000	Haiti
6,521	50	10.01	86.35	52.67	0.87	15,395	Honduras
2,863	83	16.07	698.11	17.36	24.58	19,461	Hungary
–	3,429	8.22	261.57	86.92	2,124.27	1,740,164	Iceland
436	89	0.63	15.78	13.88	18.93	5,352,303	India
88	92	4.02	64.12	8.01	16.79	4,797,060	Indonesia
1,881	55	8.77	–	28.44	253.09	419,000	Iran
–	61	2.16	–	14.69	55.13	26,789	Iraq
–	533	18.23	938.34	495.85	1,082.46	329,777	Ireland
–	360	29.92	25.66	24.17	44.70	24,661	Israel
19,288	227	8.83	227.76	76.93	121.63	540,523	Italy
–	85	19.36	42.59	23.87	0.16	12,658	Jamaica
–	440	4.89	134.00	10.26	0.03	5,935,722	Japan
2,495	52	18.71	–	8.19	95.18	1,025	Jordan
–	4	1.91	101.71	113.18	304.66	26,951	Kazakhstan
–	27	1.53	3.99	61.10	65.21	205,587	Kenya
1,860	22	0.98	52.86	102.51	382.20	198	Kirgistan
–	63	0.89	111.28	6.00	3.14	278,500	Korea (North)

: World Resources 2000–2001, Table AF.2; **Col 6:** FAO, database <http://apps.fao.org/>; **Col 7:** FAO World fisheries production by capture and culture by country, 1999 <ftp://ftp.fao.org/fi/stat/summ_99/aqua_a-0.pdf>

AGRICULTURE

Countries	1 LAND AREA *1998* 1,000 hectares	2 CROPLAND			3 AGRICULTURAL WORKERS As percentage of total workforce *2000*	4 TRACTORS Per 1,000 people *1998*
		AREA *1997* 1,000 hectares	**IRRIGATED LAND** Percentage of cropland *1997*	**CROP YIELD** kg per hectare *1996–98*		
Korea (South)	9,873	1,924	60%	6,589	10%	63.6
Kuwait	1,782	7	71%	–	1%	10.3
Laos	23,080	852	19%	2,642	76%	0.4
Latvia	6,205	1,830	1%	2,123	12%	331.5
Lebanon	1,023	308	38%	2,456	4%	124.7
Lesotho	3,035	325	1%	1,011	38%	5.9
Liberia	9,632	327	1%	1,262	68%	0.4
Libya	175,954	2,115	22%	820	6%	309.1
Lithuania	6,480	3,006	0%	2,474	12%	344.6
Luxembourg	260	–	0%	–	2%	–
Macedonia	2,543	658	8%	2,721	13%	435.5
Madagascar	58,154	3,108	35%	1,961	74%	0.6
Malawi	9,408	1,710	2%	1,224	83%	0.3
Malaysia	32,855	7,605	4%	–	19%	24.1
Maldives	–	–	0%	–	23%	–
Mali	122,019	4,650	2%	1,004	81%	0.6
Mauritania	102,522	502	10%	743	53%	0.6
Mauritius	203	–	0%	–	12%	6.1
Mexico	190,869	27,300	24%	2,653	21%	19.7
Moldova	3,297	2,183	14%	2,860	23%	84.5
Mongolia	156,650	1,320	6%	687	24%	21.0
Morocco	44,630	9,595	13%	1,215	36%	10.4
Mozambique	78,409	3,180	3%	827	81%	0.7
Namibia	82,329	820	1%	315	41%	10.7
Nepal	14,300	2,968	38%	1,959	93%	0.4
Netherlands	3,392	935	60%	7,445	3%	610.1
New Zealand	26,799	3,280	9%	5,637	9%	436.8
Nicaragua	12,140	2,746	3%	1,611	20%	6.8
Niger	126,670	5,000	1%	339	88%	0.0
Nigeria	91,077	30,738	1%	1,207	33%	2.0
Norway	30,683	902	14%	4,041	5%	1,370.4
Oman	21,246	63	98%	–	36%	0.6
Pakistan	77,088	21,600	81%	2,077	47%	11.9
Panama	7,443	655	5%	2,164	20%	20.4
Papua New Guinea	45,286	670	0%	3,840	74%	0.7
Paraguay	39,730	2,285	3%	2,365	34%	23.5
Peru	128,000	4,200	42%	2,731	30%	4.5
Philippines	29,817	9,520	16%	2,336	40%	0.9
Poland	30,442	14,424	1%	2,937	22%	296.3
Portugal	9,150	2,900	22%	2,398	13%	233.8
Qatar	1,100	–	0%	–	1%	15.0
Romania	23,034	9,900	31%	2,868	15%	96.5

Sources: Col 1: *World Resources 1998–1999;* **Col 2:** *World Resources 2000–2001, Tables AF.1 and AF.2;* **Col 3:** FAO, database <http://apps.fao.org/>;
Col 4: FAO, *Production Yearbook 2000*

5 AGROCHEMICALS		6 LIVESTOCK PRODUCTION				7 FISH	Countries
PESTICIDES kg used per hectare of cropland 1996	FERTILIZERS kg used per hectare of cropland 1995–97	CHICKENS	PIGS	BEEF AND VEAL	LAMB AND MUTTON	Total production 1999	
13,829	693	8.13	268.22	21.34	0.01	2,422,774	Korea (South)
–	855	14.44	–	5.33	1,097.18	6,535	Kuwait
57	5	2.21	151.85	32.13	–	60,403	Laos
208	30	2.16	174.10	68.46	7.20	125,857	Latvia
–	188	19.50	48.14	51.49	115.85	3,860	Lebanon
–	19	1.02	30.65	41.77	152.33	34	Lesotho
–	0	2.22	31.71	2.75	22.31	15,472	Liberia
–	36	13.52	–	18.90	557.66	32,550	Libya
312	41	5.04	232.23	152.19	9.77	35,244	Lithuania
–	–	23.60	1,038.00	78.70	–	–	Luxembourg
7,718	67	5.98	45.95	31.47	147.49	1,804	Macedonia
28	4	2.39	39.52	72.64	15.03	141,057	Madagascar
–	31	1.75	22.43	7.43	2.56	45,982	Malawi
5,982	158	26.66	205.09	7.87	0.50	1,406,895	Malaysia
–	–	–	–	–	–	133,547	Maldives
136	8	3.25	6.23	61.67	185.01	98,766	Mali
–	8	1.85	–	32.46	449.53	47,811	Mauritania
–	–	19.00	12.09	12.92	4.31	12,089	Mauritius
–	54	10.63	126.92	66.41	21.11	1,250,592	Mexico
1,434	54	2.40	123.74	32.60	58.21	1,630	Moldova
–	2	0.01	5.63	333.99	1,776.55	524	Mongolia
–	32	8.11	0.42	26.78	217.55	748,183	Morocco
–	2	1.98	10.87	13.89	3.50	35,560	Mozambique
–	0	2.17	20.28	155.16	256.12	299,196	Namibia
21	35	0.61	18.80	24.52	13.80	25,780	Nepal
11,842	821	35.42	1,190.93	141.65	46.39	623,396	Netherlands
2,215	216	18.64	197.82	874.63	8,108.18	685,734	New Zealand
357	14	4.47	29.56	69.51	0.34	24,767	Nicaragua
–	2	2.70	2.89	29.08	84.93	11,014	Niger
–	5	1.54	15.51	16.03	72.98	477,365	Nigeria
941	218	6.00	309.16	85.97	247.28	3,086,108	Norway
24,125	122	2.12	–	14.42	169.42	108,819	Oman
365	114	1.80	–	13.66	70.79	674,606	Pakistan
–	54	15.76	114.28	106.01	–	123,734	Panama
1,750	19	1.23	301.64	3.74	0.33	53,763	Papua New Guinea
1,542	13	10.30	450.05	256.55	30.08	25,095	Paraguay
–	43	8.43	71.81	35.62	97.71	8,437,565	Peru
–	82	5.88	236.44	10.72	0.12	2,198,825	Philippines
490	122	9.26	557.20	52.48	2.75	268,822	Poland
2,584	84	20.44	523.65	41.67	221.65	215,230	Portugal
–	–	4.67	–	7.08	872.57	4,207	Qatar
1,617	41	10.97	376.23	62.17	254.03	16,841	Romania

World Resources 2000–2001, Table AF.2; **Col 6:** FAO, database <http://apps.fao.org/>; **Col 7:** FAO World fisheries production by capture and culture by country, 1999 <ftp://ftp.fao.org/fi/stat/summ_99/aqua_a-0.pdf>

AGRICULTURE

Countries	1 LAND AREA *1998* 1,000 hectares	2 CROPLAND			3 AGRICULTURAL WORKERS As percentage of total workforce *2000*	4 TRACTORS Per 1,000 people *1998*
		AREA *1997* 1,000 hectares	IRRIGATED LAND Percentage of cropland *1997*	CROP YIELD kg per hectare *1996–98*		
Russia	1,688,850	127,962	4%	1,295	10%	101.4
Rwanda	2,467	1,150	0%	1,188	90%	0.0
Samoa	–	–	0%	–	–	2.0
Saudi Arabia	214,969	3,830	42%	3,880	10%	12.9
Senegal	19,253	2,266	3%	719	74%	0.2
Seychelles	46		0%	–	–	
Sierra Leone	7,162	546	5%	1,223	62%	0.1
Singapore	61	1	0%	–	0%	
Slovakia	4,808	1,605	12%	4,159	9%	91.8
Slovenia	2,012	285	1%	5,357	2%	4,761.4
Solomon Islands	2,799	60	0%	–	73%	–
Somalia	62,734	1,061	19%	410	71%	0.6
South Africa	122,104	16,300	8%	2,220	10%	54.3
Spain	49,944	19,164	19%	3,156	7%	627.4
Sri Lanka	6,463	1,888	32%	3,103	46%	2.0
Sudan	237,600	16,900	12%	602	61%	1.5
Suriname	15,600	67	90%	3,746	19%	44.3
Swaziland	1,720	–	0%	–	34%	24.3
Sweden	41,162	2,799	4%	4,707	3%	1,051.0
Switzerland	3,955	444	6%	6,709	4%	658.8
Syria	18,378	5,521	21%	1,543	28%	66.7
Tajikistan	14,060	890	81%	1,682	34%	37.1
Tanzania	88,359	4,000	4%	1,261	80%	0.6
Thailand	51,089	20,445	25%	2,481	56%	10.5
Togo	5,439	2,430	0%	876	60%	0.1
Trinidad & Tobago	513	122	18%	3,628	9%	52.9
Tunisia	15,536	4,900	8%	1,240	25%	37.5
Turkey	76,963	29,162	14%	2,196	46%	59.8
Turkmenistan	46,993	1,695	106%	1,138	33%	78.7
Uganda	19,965	6,810	0%	1,248	80%	0.6
Ukraine	57,935	34,081	7%	2,214	14%	91.0
United Arab Emirates	8,360	81	89%	–	5%	4.6
United Kingdom	24,160	6,425	2%	6,880	2%	929.4
United States of America	915,912	179,000	12%	5,352	2%	1,578.90
Uruguay	17,481	1,307	11%	3,370	13%	173.7
Uzbekistan	41,424	4,850	88%	2,279	28%	59.5
Venezuela	88,205	3,490	6%	2,952	8%	60.5
Vietnam	32,549	7,202	32%	3,760	67%	4.4
Yemen	52,797	1,555	31%	974	51%	2.0
Yugoslavia	10,200	4,058	2%	3,797	20%	402.9
Zambia	74,339	5,265	1%	1,585	69%	2.3
Zimbabwe	38,685	3,210	5%	1,283	63%	6.6

Sources: Col 1: *World Resources 1998–1999;* **Col 2:** *World Resources 2000–2001, Tables AF.1 and AF.2;* **Col 3:** FAO, database <http://apps.fao.org/>;
Col 4: FAO, *Production Yearbook 2000*

5 AGROCHEMICALS		6 LIVESTOCK PRODUCTION				7 FISH	Countries
PESTICIDES kg used per hectare of cropland 1996	FERTILIZERS kg used per hectare of cropland 1995–97	CHICKENS	PIGS	BEEF AND VEAL	LAMB AND MUTTON	Total production 1999	
407	17	3.95	105.54	84.13	40.55	4,209,772	Russia
260	0	0.22	6.21	24.97	7.89	6,733	Rwanda
–	–	1.78	572.22	42.77	–	9,750	Samoa
–	98	19.35	–	5.11	196.60	51,949	Saudi Arabia
183	12	6.76	26.05	42.35	115.70	418,280	Senegal
–	–	12.34	285.71	1.75	–	37,992	Seychelles
–	6	2.06	8.14	16.57	24.52	59,437	Sierra Leone
–	3,247	12.78	362.09	0.06	3.57	9,081	Singapore
4,148	77	14.42	415.37	32.67	32.24	2,263	Slovakia
6,389	258	26.99	431.61	87.42	35.71	3,215	Slovenia
–	0	–	112.61	6.26	–	82,347	Solomon Islands
–	0	0.42	0.24	61.52	330.37	20,250	Somalia
57	51	9.11	49.53	64.42	196.26	592,144	South Africa
–	123	14.13	921.65	60.61	513.69	1,485,038	Spain
6,261	111	2.52	1.22	9.56	–	279,900	Sri Lanka
106	4	1.02	–	83.61	287.83	50,500	Sudan
4,877	93	4.80	59.95	34.77	6.00	13,066	Suriname
–	–	6.45	33.73	73.51	43.24	131	Swaziland
509	113	8.19	426.21	59.82	22.86	357,317	Sweden
4,576	258	5.28	357.43	91.35	39.59	2,975	Switzerland
–	68	6.95	–	26.44	631.90	14,024	Syria
–	65	0.32	0.76	9.36	169.21	80	Tajikistan
–	9	1.31	7.31	59.37	27.91	310,270	Tanzania
1,116	75	13.62	138.72	13.53	0.26	3,607,707	Thailand
95	6	2.53	96.13	9.21	48.60	23,074	Togo
11,827	203	13.13	27.80	4.48	3.48	15,027	Trinidad & Tobago
–	25	6.26	0.31	28.54	433.45	93,170	Tunisia
1,145	66	8.03	0.06	31.50	296.99	638,097	Turkey
6,744	82	0.84	2.47	76.00	823.31	9,292	Turkmenistan
17	0	1.43	41.42	27.68	15.67	226,457	Uganda
2,001	26	2.70	155.56	115.01	23.66	441,672	Ukraine
–	421	10.22	–	23.41	498.85	117,607	United Arab Emirates
4,745	343	13.50	214.88	40.80	308.23	992,559	United Kingdom
1,599	151	32.33	351.80	133.10	12.43	5,228,324	United States of America
1,316	89	15.58	106.68	584.36	943.96	103,043	Uruguay
–	109	0.58	10.69	95.98	195.29	8,536	Uzbekistan
1,403	86	13.24	78.61	68.27	6.62	422,607	Venezuela
–	206	2.61	263.32	6.92	–	1,794,910	Vietnam
–	9	3.86	–	30.96	130.91	123,252	Yemen
887	44	7.42	776.50	43.59	136.75	9,940	Yugoslavia
317	10	3.82	26.61	19.10	3.45	71,507	Zambia
531	59	1.80	20.14	35.64	2.85	12,591	Zimbabwe

World Resources 2000–2001, Table AF.2; **Col 6:** FAO, database <http://apps.fao.org/>; **Col 7:** FAO *World fisheries production by capture and _lture by country*, 1999 <ftp://ftp.fao.org/fi/stat/summ_99/aqua_a-0.pdf>

CONSUMPTION

Countries	1 POPULATION		2 CALORIES	3 ANIMAL PRODUCTS	4 ANIMAL FEED	5 WATER	
	Total 2000 thousands	Projected annual increase 1998—2015	Average daily calorie supply per person 1997	Average daily consumption per person 1997 kilo calories	Percentage of grain consumed used for animal feed 1994	Annual renewable water available per person 2000 cubic meters	Percentage populatio without ac to safe wa 1997
Afghanistan	21,765	4.8%	1,745	138	0	2,986	€
Albania	3,134	-0.5%	2,961	682	0	13,592	
Algeria	30,291	2.0%	2,853	265	22	472	2:
Angola	13,134	3.2%	1,903	138	0	14,009	6!
Argentina	37,032	1.3%	3,093	886	51	26,545	2!
Armenia	3,787	0.7%	2,371	404	19	2,799	
Australia	19,138	1.3%	3,224	1,021	60	18,393	
Austria	8,080	0.4%	3,536	1,256	69	10,396	
Azerbaijan	8,041	1.1%	2,236	367	28	3,619	
Bahamas	304	1.8%	–	–	–	–	●
Bahrain	640	2.7%	–	–	–	–	
Bangladesh	137,439	2.2%	2,086	69	0	8,808	!
Belarus	10,187	-0.1%	3,226	919	70	5,694	
Belgium	10,249	0.3%	3,619	1,151	44	1,561	
Belize	226	2.0%	2,907	648	–	70,695	1
Benin	6,272	3.0%	2,487	104	0	4,114	4
Bhutan	2,085	2.1%	–	–	0	45,561	4:
Bolivia	8,329	2.4%	2,174	406	34	38,806	3
Bosnia-Herzegovina	3,977	-0.8%	2,266	228	–	9,429	●
Botswana	1,541	2.2%	2,183	396	0	9,538	1●
Brazil	170,406	1.4%	2,974	585	60	42,944	2
Brunei	328	2.5%	–	–	–	–	
Bulgaria	7,949	-0.9%	2,686	617	55	2,290	
Burkina Faso	11,535	2.5%	2,121	109	2	1,690	5
Burma	47,749	1.7%	2,862	119	0	21,898	4●
Burundi	6,356	1.2%	1,685	44	0	566	4
Cambodia	13,104	3.1%	2,048	163	0	36,340	7●
Cameroon	14,876	2.5%	2,111	125	0	18,016	5●
Canada	30,757	1.1%	3,119	843	76	90,777	
Central African Rep.	3,717	2.4%	2,016	190	0	37,931	6.
Chad	7,885	3.1%	2,032	113	0	5,453	7
Chile	15,211	1.5%	2,796	612	34	61,007	
China	1,282,437	1.0%	2,897	510	23	2,206	3.
Colombia	42,105	1.9%	2,597	428	30	50,659	1
Comoros	706	3.0%	–	–	–	–	4
Congo	3,018	3.1%	2,144	144	0	16,328	6
Congo, Dem. Rep.	50,948	3.3%	1,755	47	0	413,461	5
Costa Rica	4,024	2.8%	2,649	469	51	27,936	
Côte d'Ivoire	16,013	2.4%	2,610	97	3	4,790	5
Croatia	4,654	0.3%	2,445	459	–	15,342	
Cuba	11,199	0.5%	2,480	301	0	3,393	
Cyprus	784	1.4%	–	–	–	–	

Sources: Col 1: World Health Organization, 2002; **Col 2:** World Resources 2000—2001, Table AF.3; **Col 3:** World Resources 2000—2001, Table AF.3; **Col 4:** World Resources 1996—97, Table10.3; **Col 5:** Population Action International, 2002; Human Development Report 1999

6 UNDERNUTRITION	7 OVERNUTRITION			8 FOOD AID	9 CONSUMER ACTION	Countries
Percentage of children underweight 1990—97	Death rates from coronary heart disease per 100,000 people latest available data		Percentage of people with diabetes 2000	Tons of food from World Food Program 2000	Number of organizations affiliated to Consumer Action International 2002	
	men	women				
48	–	–	1.5%	188,147	–	Afghanistan
–	134.68	50.51	2.2%	–	1	Albania
13	–	–	2.0%	13,251	–	Algeria
42	–	–	0.4%	146,821	–	Angola
–	127.66	35.13	3.6%	–	5	Argentina
–	439.68	184.53	2.1%	12,254	2	Armenia
–	164.30	57.07	4.3%	–	10	Australia
–	197.57	60.49	4.0%	–	1	Austria
10	665.45	299.74	1.8%	8,937	1	Azerbaijan
–	–	–	4.6%	–	–	Bahamas
–	271.39	160.04	7.2%	–	–	Bahrain
56	–	–	1.5%	89,239	2	Bangladesh
–	709.06	253.52	3.0%	–	2	Belarus
–	137.06	44.12	3.5%	–	5	Belgium
6	134.25	111.50	3.3%	86	–	Belize
29	–	–	0.5%	3,090	3	Benin
38	–	–	1.4%	3,548	–	Bhutan
16	–	–	2.1%	20,025	1	Bolivia
–	–	–	3.8%	–	–	Bosnia-Herzegovina
17	0.00	0.00	1.1%	–	1	Botswana
6	415.42	205.80	2.9%	–	4	Brazil
–	–	–	4.8%	–	–	Brunei
–	–	–	3.3%	–	1	Bulgaria
30	301.89	111.17	0.5%	1,141	1	Burkina Faso
43	0.39	1.00	0.9%	–	–	Burma
37	–	–	0.5%	11,100	1	Burundi
52	–	–	0.7%	59,664	–	Cambodia
14	–	–	0.4%	362	1	Cameroon
–	165.96	56.84	4.4%	–	8	Canada
27	–	–	0.5%	2,316	–	Central African Rep.
39	–	–	0.4%	7,837	1	Chad
1	122.60	49.71	3.5%	–	3	Chile
16	71.22	48.02	1.6%	84,300	1	China
8	173.03	100.02	2.6%	20,248	2	Colombia
–	–	–	0.5%	–	–	Comoros
17	–	–	0.4%	8,469	1	Congo
34	–	–	0.4%	30,856	–	Congo, Dem. Rep.
2	152.80	81.40	4.5%	–	1	Costa Rica
24	–	–	0.5%	1,520	1	Côte d'Ivoire
1	266.14	97.04	3.8%	–	1	Croatia
9	222.83	131.68	5.4%	2,788	1	Cuba
–	–	–	4.4%	–	1	Cyprus

World Resources 2000—2001, Table AF.3; **Col 7:** World Health Organization Statistical Information System 2001, WHO Collaborating Centre in the ...n of Cardiology, University of Ottawa, 2001; A F Amos, D J McCarty, P Zimmet, The Rising Global Burden of Diabetes Diabetic Medicine 14: S7-S85, **Col 8:** World Food Programme <www.wfp.org>; **Col 9:** Consumer International <www.consumersinternational.org/members/index.html>

CONSUMPTION

Countries	1 POPULATION		2 CALORIES	3 ANIMAL PRODUCTS	4 ANIMAL FEED	5 WATER	
	Total 2000 thousands	Projected annual increase 1998—2015	Average daily calorie supply per person 1997	Average daily consumption per person 1997 kilo calories	Percentage of grain consumed used for animal feed 1994	Annual renewable water available per person 2000 cubic meters	Percentage population without access to safe water 1997
Czech Republic	10,272	0.0%	3,244	816	62	1,558	
Denmark	5,320	0.3%	3,407	1,259	81	1,116	
Djibouti	632	2.3%	–	–	–	–	
Dominican Republic	8,373	1.7%	2,288	339	58	2,508	
Ecuador	12,646	2.1%	2,679	471	38	34,952	
Egypt	67,884	1.9%	3,287	221	35	1,009	
El Salvador	6,278	2.1%	2,562	297	24	2,819	
Equatorial Guinea	457	2.6%	–	–	–	65,688	
Eritrea	3,659	1.7%	1,622	92	–	2,405	
Estonia	1,393	-1.2%	2,849	797	64	9,186	
Ethiopia	62,908	2.8%	1,858	101	0	1,749	
Fiji	814	1.2%	2,865	589	0	–	
Finland	5,172	0.4%	3,100	1,195	60	21,269	
France	59,238	0.4%	3,518	1,334	65	2,870	
Gabon	1,230	2.8%	2,556	327	0	133,324	
Gambia	1,303	3.4%	2,350	126	0	6,141	
Georgia	5,262	-0.4%	2,614	316	21	12,638	
Germany	82,017	0.3%	3,382	1,050	60	2,170	
Ghana	19,306	2.5%	2,611	84	3	2,756	
Greece	10,610	0.4%	3,649	798	52	6,786	
Guatemala	11,385	2.7%	2,339	205	25	11,805	
Guinea	8,154	2.9%	2,232	62	0	27,716	
Guinea-Bissau	1,199	2.4%	2,430	165	0	22,515	
Guyana	761	0.4%	2,530	343	5	316,891	
Haiti	8,142	1.7%	1,869	99	2	1,486	
Honduras	6,417	2.8%	2,403	332	47	14,977	
Hungary	9,968	-0.4%	3,313	1,046	75	19,062	
Iceland	279	0.9%	3,117	1,224	–	608,684	
India	1,008,937	1.8%	2,496	174	3	1,891	
Indonesia	212,092	1.5%	2,886	134	9	13,381	
Iran	70,330	1.9%	2,836	279	18	1,821	
Iraq	22,946	2.9%	2,619	95	23	4,842	
Ireland	3,803	0.8%	3,565	1,126	63	12,095	
Israel	6,040	3.0%	3,278	589	60	464	
Italy	57,530	0.1%	3,507	902	48	3,042	
Jamaica	2,576	0.8%	2,553	455	32	3,649	
Japan	127,096	0.3%	2,932	598	46	3,423	
Jordan	4,913	4.2%	3,014	266	54	142	
Kazakhstan	16,172	-0.3%	3,085	609	54	6,777	
Kenya	30,669	2.7%	1,977	241	2	985	
Kirgistan	4,921	1.1%	2,447	543	46	9,450	
Korea (North)	22.268	1.1%	1,837	150	0	3,462	

Sources: Col 1: World Health Organization, 2002; Col 2: World Resources 2000—2001, Table AF.3; Col 3: World Resources 2000—2001, Table AF.3; Col 4: World Resources 1996—97, Table10.3; Col 5: Population Action International, 2002; Human Development Report 1999

6 UNDERNUTRITION	7 OVERNUTRITION			8 FOOD AID	9 CONSUMER ACTION	Countries
Percentage of children underweight 1990—97	Death rates from coronary heart disease per 100,000 people latest available data		Percentage of people with diabetes 2000	Tons of food from World Food Program 2000	Number of organizations affiliated to Consumer Action International 2002	
	men	women				
1	398.91	143.45	3.0%	–	2	Czech Republic
–	–	–	4.2%	–	1	Denmark
–	–	–	1.3%	8,019	–	Djibouti
6	114.56	68.43	3.2%	1,583	2	Dominican Republic
17	38.86	19.82	2.3%	12,711	1	Ecuador
15	87.92	39.61	5.8%	30,552	1	Egypt
11	75.51	48.31	3.9%	2,100	1	El Salvador
–	–	–	0.5%	–	–	Equatorial Guinea
44	–	–	0.6%	150,399	–	Eritrea
–	538.51	179.81	4.5%	–	–	Estonia
48	–	–	0.5%	604,717	–	Ethiopia
8	–	–	5.8%	–	2	Fiji
–	300.40	79.61	5.3%	–	6	Finland
–	77.91	17.70	3.4%	–	4	France
–	–	–	0.6%	1,100	1	Gabon
26	–	–	0.6%	2,542	–	Gambia
–	590.62	261.73	2.8%	16,595	2	Georgia
–	184.05	62.28	4.3%	–	3	Germany
27	–	–	0.6%	3,621	1	Ghana
–	155.08	43.68	5.0%	–	2	Greece
27	60.80	36.26	3.5%	5,163	1	Guatemala
–	–	–	0.5%	1,548	–	Guinea
23	–	–	0.6%	1,807	–	Guinea-Bissau
12	203.65	108.20	3.4%	–	1	Guyana
28	–	–	1.1%	11,019	–	Haiti
18	–	–	3.4%	1,921	1	Honduras
2	426.54	151.37	3.1%	–	2	Hungary
–	189.98	49.41	4.0%	–	1	Iceland
53	–	–	2.7%	79,853	18	India
34	–	–	3.4%	148,336	2	Indonesia
16	–	–	1.5%	3,217	–	Iran
23	–	–	1.6%	2,890	–	Iraq
–	296.97	92.93	2.9%	–	2	Ireland
–	128.50	55.75	4.4%	–	4	Israel
–	114.42	32.11	5.1%	–	3	Italy
10	85.24	53.87	7.4%	–	2	Jamaica
–	40.34	14.63	5.7%	17,856	6	Japan
9	–	–	3.2%	–	1	Jordan
8	648.97	247.37	2.1%	–	1	Kazakhstan
23	–	–	0.5%	274,437	1	Kenya
–	441.74	203.32	1.6%	–	–	Kirgistan
–	–	–	2.0%	452,945	–	Korea (North)

World Resources 2000—2001, Table AF.3; **Col 7:** World Health Organization Statistical Information System 2001, WHO Collaborating Centre in the on of Cardiology, University of Ottawa, 2001; A F Amos, D J McCarty, P Zimmet, The Rising Global Burden of Diabetes Diabetic Medicine 14: S7-S85, **Col 8:** World Food Programme <www.wfp.org>; **Col 9:** Consumer International <www.consumersinternational.org/members/index.html>

CONSUMPTION

Countries	POPULATION		CALORIES	ANIMAL PRODUCTS	ANIMAL FEED	WATER	
	1		2	3	4	5	
	Total 2000 thousands	Projected annual increase 1998—2015	Average daily calorie supply per person 1997	Average daily consumption per person 1997 kilo calories	Percentage of grain consumed used for animal feed 1994	Annual renewable water available per person 2000 cubic meters	Percentag populati without ac to safe w 1997
Korea (South)	46,740	0.9%	3,155	502	50	1,493	
Kuwait	1,914	-1.1%	3,096	728	25	0	
Laos	5,279	2.5%	2,108	136	0	63,180	5
Latvia	2,421	-1.0%	2,864	705	64	14,625	
Lebanon	3,496	2.6%	3,277	448	35	1,373	
Lesotho	2,035	1.9%	2,244	136	25	2,556	3
Liberia	2,913	3.1%	2,044	65	0	79,641	
Libya	5,290	2.1%	3,289	341	27	151	
Lithuania	3,696	-0.1%	3,261	794	63	6,737	
Luxembourg	437	1.4%	–	–	–	3,663	
Macedonia	2, 034	0.6%	2,664	488	–	3,442	
Madagascar	15,970	2.9%	2,022	194	0	21,102	7
Malawi	11,308	1.8%	2,043	56	2	1,645	5
Malaysia	22,218	2.2%	2,977	570	46	26,104	2
Maldives	291	3.0%	–	–	–	–	4
Mali	11,351	2.6%	2,030	214	2	8,810	3
Mauritania	2,665	2.9%	2,622	443	0	4,278	2
Mauritius	1,161	0.9%	–	–	0	–	
Mexico	98,872	1.7%	3,097	520	38	4,673	1
Moldova	4,295	-0.2%	2,567	388	50	2,724	
Mongolia	2,533	1.3%	1,917	845	0	13,737	6
Morocco	29,878	2.0%	3,078	210	26	1,004	3
Mozambique	18,292	3.0%	1,832	44	0	11,535	3
Namibia	1,757	2.5%	2,183	272	–	25,902	1
Nepal	23,043	2.4%	2,366	152	0	9,122	2
Netherlands	15,864	0.6%	3,284	1,135	40	5,736	
New Zealand	3,778	1.2%	3,395	1,303	47	86,554	
Nicaragua	5,071	2.9%	2,186	165	5	37,504	3
Niger	10,832	3.5%	2,097	113	0	3,000	5
Nigeria	113,862	2.9%	2,735	103	1	2,459	5
Norway	4,469	0.5%	3,357	1,124	66	87,939	
Oman	2,538	3.6%	–	–	0	394	1
Pakistan	141,256	2.6%	2,476	370	4	1,805	2
Panama	2,856	1.8%	2,430	537	39	51,616	
Papua New Guinea	4,809	2.5%	2,224	247	0	166,555	6
Paraguay	5,496	2.7%	2,566	594	2	17,102	4
Peru	25,662	1.8%	2,302	339	38	73,651	3
Philippines	75,653	2.2%	2,366	364	25	6,332	1
Poland	38,605	0.1%	3,366	884	61	1,632	
Portugal	10,016	0.1%	3,667	995	58	7,189	
Qatar	565	2.2%	–	–	–	–	
Romania	22,438	-0.3%	3,253	697	65	9,760	

Sources: **Col 1:** World Health Organization, 2002; **Col 2:** World Resources 2000—2001, Table AF.3; **Col 3:** World Resources 2000—2001, Table AF.3; **Col 4:** World Resources 1996—97, Table10.3; **Col 5:** Population Action International, 2002; Human Development Report 1999

6 UNDERNUTRITION	7 OVERNUTRITION			8 FOOD AID	9 CONSUMER ACTION	Countries
Percentage of children underweight 1990—97	Death rates from coronary heart disease per 100,000 people latest available data		Percentage of people with diabetes 2000	Tons of food from World Food Program 2000	Number of organizations affiliated to Consumer Action International 2002	
	men	women				
–	35.22	12.63	4.5%	–	5	Korea (South)
6	211.23	157.39	6.3%	–	–	Kuwait
40	–	–	0.7%	3,730	–	Laos
–	718.43	219.06	3.6%	–	2	Latvia
3	–	–	4.2%	–	–	Lebanon
16	–	–	1.1%	3,098	2	Lesotho
–	–	–	0.7%	68,576	2	Liberia
5	–	–	2.7%	–	–	Libya
–	449.49	142.02	3.5%	–	2	Lithuania
–	137.55	36.88	4.2%	–	2	Luxembourg
–	223.81	90.35	3.1%	–	1	Macedonia
40	–	–	0.5%	8,923	–	Madagascar
30	–	–	0.5%	10,478	2	Malawi
19	–	–	4.2%	–	6	Malaysia
–	–	–	1.2%	–	–	Maldives
40	–	–	0.5%	10,184	–	Mali
23	–	–	0.6%	3,411	–	Mauritania
–	–	–	8.2%	–	2	Mauritius
14	125.34	66.13	4.1%	–	2	Mexico
–	697.70	437.26	2.5%	–	–	Moldova
10	176.23	94.81	0.9%	–	1	Mongolia
9	–	–	2.1%	9,752	2	Morocco
27	–	–	0.6%	68,390	1	Mozambique
26	–	–	1.3%	1,932	1	Namibia
47	–	–	1.4%	33,723	2	Nepal
–	167.31	54.68	3.9%	–	2	Netherlands
–	200.24	62.37	4.1%	–	3	New Zealand
12	99.36	65.85	3.3%	22,244	1	Nicaragua
43	–	–	0.4%	8,306	–	Niger
36	–	–	0.6%	–	2	Nigeria
–	231.92	67.65	3.9%	–	2	Norway
23	–	–	5.1%	–	–	Oman
38	–	–	5.1%	12,354	1	Pakistan
7	117.39	71.92	4.6%	–	4	Panama
30	–	–	0.8%	–	1	Papua New Guinea
4	259.30	128.02	2.0%	–	–	Paraguay
8	38.21	18.09	2.6%	13,953	2	Peru
28	147.49	67.34	3.3%	–	4	Philippines
–	246.41	70.09	2.8%	–	2	Poland
–	110.01	35.00	4.7%	–	2	Portugal
–	–	–	8.9%	–	–	Qatar
6	329.60	150.51	3.0%	–	2	Romania

: World Resources 2000—2001, Table AF.3; **Col 7:** World Health Organization Statistical Information System 2001, WHO Collaborating Centre in the
ion of Cardiology, University of Ottawa, 2001; A F Amos, D J McCarty, P Zimmet, The Rising Global Burden of Diabetes Diabetic Medicine 14: S7-S85,
: **Col 8:** World Food Programme <www.wfp.org>; **Col 9:** Consumer International <www.consumersinternational.org/members/index.html>

CONSUMPTION

Countries	1 POPULATION Total 2000 thousands	1 POPULATION Projected annual increase 1998—2015	2 CALORIES Average daily calorie supply per person 1997	3 ANIMAL PRODUCTS Average daily consumption per person 1997 kilo calories	4 ANIMAL FEED Percentage of grain consumed used for animal feed 1994	5 WATER Annual renewable water available per person 2000 cubic meters	5 WATER Percentage population without ac to safe wa 1997
Russia	145,491	-0.2%	2,904	726	57	29,163	
Rwanda	7,609	1.2%	2,057	60	7	828	
Samoa	159	-0.1%	–	–	–	–	3
Saudi Arabia	20,346	2.8%	2,783	415	74	118	
Senegal	9,421	2.5%	2,418	193	0	4,182	3
–Seychelles	80	1.5%	–	–	–	–	
Sierra Leone	4,405	0.8%	2,035	66	2	36,325	6
Singapore	4,018	2.9%	–	–	20	–	
Slovakia	5,399	0.3%	2,984	786	64	15,374	
Slovenia	1,988	0.4%	3,101	841	–	9,307	
Solomon Islands	447	3.5%	2,122	197	–	–	3
Somalia	8,778	2.1%	1,566	737	4	1,789	
South Africa	43,309	1.8%	2,990	409	33	1,154	1
Spain	39,910	0.2%	3,310	860	68	2,809	
Sri Lanka	18,924	1.1%	2,302	143	0	2,642	4
Sudan	31,095	2.3%	2,395	473	0	4,953	2
Suriname	417	0.4%	2,665	332	0	479,434	
Swaziland	925	1.9%	–	–	0	–	5
Sweden	8,842	0.3%	3,194	1,075	75	20,131	
Switzerland	7,170	0.5%	3,223	1,119	55	7,391	
Syria	16,189	2.7%	3,352	399	29	2,761	1
Tajikistan	6,087	1.4%	2,001	143	17	13,077	
Tanzania	35,119	3.0%	1,995	129	3	2,534	3
Thailand	62,806	1.4%	2,360	282	30	6,526	1
Togo	4,527	2.7%	2,469	102	21	2,651	4
Trinidad & Tobago	1,294	0.6%	2,661	382	35	–	
Tunisia	9,459	1.5%	3,283	279	29	412	
Turkey	66,668	1.7%	3,525	397	31	3,510	5
Turkmenistan	4,737	2.6%	2,306	417	46	12,856	
Uganda	23,300	3.1%	2,085	138	0	2,833	5
Ukraine	49,568	-0.5%	2,795	583	59	2,816	
United Arab Emirates	2,606	2.6%	3,390	827	20	77	
United Kingdom	59,415	0.3%	3,276	1,024	50	2,474	
United States of America	283,230	1.1%	3,699	995	68	8,749	
Uruguay	3,337	0.7%	2,816	944	13	39,855	
Uzbekistan	24,881	1.9%	2,433	432	28	4,598	
Venezuela	24,170	2.2%	2,321	361	33	35,002	2
Vietnam	78,137	1.7%	2,484	226	0	11,406	5
Yemen	18,349	4.7%	2,051	144	0	223	3
Yugoslavia	10,552	0.4%	3,031	1,041	52	17,816	6
Zambia	10,421	2.6%	1,970	113	4	11,131	2
Zimbabwe	12,627	2.1%	2,145	180	9	1,117	

 Sources: Col 1: World Health Organization, 2002; **Col 2:** World Resources 2000—2001, Table AF.3; **Col 3:** World Resources 2000—2001, Table AF.3; **Col 4:** World Resources 1996—97, Table10.3; **Col 5:** Population Action International, 2002; Human Development Report 1999

6 ERNUTRITION	7 OVERNUTRITION			8 FOOD AID	9 CONSUMER ACTION	Countries
ercentage of children underweight 1990—97	Death rates from coronary heart disease per 100,000 people latest available data		Percentage of people with diabetes 2000	Tons of food from World Food Program 2000	Number of organizations affiliated to Consumer Action International 2002	
	men	women				
3	589.27	202.03	2.9%	32,791	4	Russia
27	–	–	0.6%	139,545	–	Rwanda
–	–	–	3.8%	–	–	Samoa
–	–	–	6.3%	–	–	Saudi Arabia
22	–	–	0.5%	25,217	4	Senegal–
–	–	–	0.0%	–	1	Seychelles
29	–	–	0.6%	3,049	–	Sierra Leone
–	230.88	106.56	7.4%	–	–	Singapore
–	410.85	151.46	2.7%	–	1	Slovakia
–	154.13	51.73	3.7%	–	1	Slovenia
21	–	–	0.7%	–	–	Solomon Islands
–	–	–	0.5%	20,395	–	Somalia
9	94.61	43.53	2.5%	–	4	South Africa
–	110.01	28.25	4.6%	–	13	Spain
34	89.52	24.17	1.8%	12,292	2	Sri Lanka
34	–	–	1.5%	116,231	–	Sudan
–	250.47	125.78	3.3%	–	–	Suriname
–	–	–	1.1%	–	–	Swaziland
–	189.62	57.55	5.1%	–	4	Sweden
–	–	–	4.1%	–	2	Switzerland
13	100.81	40.70	3.1%	21,837	–	Syria
–	385.82	229.11	1.4%	31,067	–	Tajikistan
27	–	–	0.6%	14,463	1	Tanzania
19	5.16	2.46	2.0%	–	5	Thailand
19	–	–	0.5%	–	2	Togo
7	–	–	5.5%	–	–	Trinidad & Tobago
9	–	–	2.5%	–	1	Tunisia
10	8.80	2.45	4.6%	–	3	Turkey
–	595.49	299.32	1.5%	–	1	Turkmenistan
26	–	–	0.4%	48,820	1	Uganda
–	657.45	287.68	3.1%	–	1	Ukraine
14	–	–	8.2%	–	1	United Arab Emirates
–	236.20	83.89	3.5%	–	6	United Kingdom
1	192.49	77.23	4.8%	–	5	United States of America
5	181.20	65.65	3.7%	–	4	Uruguay
19	560.79	330.60	1.5%	–	–	Uzbekistan
5	227.05	111.53	2.8%	1,690	2	Venezuela
41	–	–	0.9%	37,717	1	Vietnam
39	–	–	1.2%	20,461	–	Yemen
2	184.34	66.98	3.6%	130,128	–	Yugoslavia
24	–	–	0.5%	19,724	2	Zambia
16	15.30	6.08	0.5%	–	2	Zimbabwe

World Resources 2000—2001, Table AF.3; **Col 7:** World Health Organization Statistical Information System 2001, WHO Collaborating Centre in the
ion of Cardiology, University of Ottawa, 2001; A F Amos, D J McCarty, P Zimmet, The Rising Global Burden of Diabetes Diabetic Medicine 14: S7-S85,
Col 8: World Food Programme <www.wfp.org>; **Col 9:** Consumer International <www.consumersinternational.org/members/index.html>

REFERENCES

Part 1: Contemporary Challenges

12—13 FEEDING THE WORLD

The State of Food and Agriculture 2001, FAO, Rome 2001
  <www.fao.org>
<www.jhuccp.org/pr/m13/m13chap2.stm#contents>
  which cites a range of FAO sources
<www.hungerhurts.org>

Undernourishment around the world
<www.fao.org>

Calories available and Grain production
World Resources 2000-2001, Table AF.3 <www.wri.org>

Over- and under-nutrition
<www.who.int/nut/db_bmi.htm>

14 — 15 POPULATION AND PRODUCTIVITY

UN Population Division, 2000 <www.undp.org>
Population Action International
  <www.populationaction.org>

Population increase
WHO 2002

Productivity
FAO statistics 1999, cited in Sustainable Agriculture in a
  Globalized Economy, ILO, Geneva 2000 <www.ilo.org>

16 — 17 ENVIRONMENTAL CHALLENGES

Lester Brown, World Grain Harvest Falling Short by 54
  Million Tons. Water Shortages Contributing to Shortfall.
  Earth Policy Institute, 2001 <www.earth-policy.org>
The State of Food and Agriculture 2001, FAO, Rome,
  Italy.
World population prospects: The 2000 Revision, United
  Nations Population Division, Department of Economic
  and Social Affairs, 2001, UN <www.undp.org>
J E Sheehy, The Consequences of Global Warming,
  International Rice Research Institute, 2000
  <www.irri.org/ar2001/sheehy2.pdf 2.12.01>
D J Farrell, Recipe for disaster? Where do we find the
  ingredients to feed our livestock? , The Food and
  Environment Tightrope, ed H Cadman, Australian Centre
  for International Agricultural Research, Canberra, ACT,
  Australia, 2000, pp96—121

Soil degradation
S Wood, K Sebastian, and S J Scherr, Pilot Analysis of
  Global Ecosystems Agroecosystems, International Food
  Policy Research Institute, World Resources Institute,
  2000. Table 15
  <www.ifpri.cgiar.org/pubs/books/page.htm 26.11.01>
Global Environment Outlook, UNEP 2002

Global Climate Change
NASA/Goddard Institute for Space Studies

18 — 19 WATER PRESSURE

Water shortage
Population Action International, based on water data from
  World Resources Institute

Irrigated land
World Resources 2000—2001, Table AF.2 <www.wri.org>

Drought
African Development Indicators 1998/99, World Bank,
  Washington DC, Table 8—14, Figure 8-5. Sourced from
  Sustainable Agriculture in a Globalised Economy,
  International Labour Organisation, 2000 <www.ilo.org>

Water shortfall
Association of California Water Agencies, California
  Water Plan, February 2000
  <www.acwanet.com/generalinfo/waterfacts/calwaterplan.a
  sp> 3.07 acre feet = 1 million gallons

20 — 21 CONSUMING DISEASE

Unsafe drinking water
Human Development Report, 1999

Foodborne infections
Europe data: WHO Surveillance Programme for Control
  of Foodborne Infections and Intoxications in Europe 7th
  Report <www.bgvv.de>
North America data: Health Canada. FoodNet MMWR.
  50(13):241-6

The tip of the iceberg
Wheeler et al Study of infectious intestinal disease in
  England: Rate in the community, presenting to general
  practice, and reported to national surveillance BMJ
  318:1046-1050, 1999

Disease agents
<www.inppaz.org.ar>
C Tirado and K Schmidt, WHO Surveillance Programme
  for Control of Foodborne Infections and Intoxicants:
  Preliminary Results and Trends Across Greater Europe ,
  Journal of Infection, 43, 2001, pp 80—84

22 — 23 UNDER-NUTRITION

Vitamin A Deficiency
<www.who.int/vaccines-
  diseases/en/vitamina/advocacy/adv05.shtml>
<whqlibdoc.who.int/hq/1996/WHO_NUT_96.10.pdf>

Iodine deficiency disorders
<www.who.int/nut/publications.htm#idd>

24 — 25 OVER-NUTRITION

Diabetes
A F Amos, D J McCarty, P Zimmet , The rising global
  burden of diabetes and its complications: estimates and
  projections to the year 2010 . Diabetic Medicine 14: S7-
  S85, 1997

Coronary Heart Disease
World Health Organisation (2001) WHO Statistical

Information System <www.who.ch>
WHO Collaborating Centre in the Division of Cardiology, University of Ottawa (2001) Global Cardiovascular Infobase <cvdinfobase.ic.gc.ca/>
Obesity
Coronary Heart Disease Statistics: Diabetes Supplement, <www.dphpc.ox.ac.uk> Sourced from Professor Boyd Swinburn, Deakin University, Victoria, Australia

26 — 27 FOOD AID
World Food Program <www.wfp.org>
Map symbol: The State of Food and Agriculture 2001, FAO, Rome, 2001, Map 2, page 20

28 — 29 FOOD AID AS POWER
FAO/WHO International Conference on Nutrition, Rome 1992.
The State of Food and Agriculture 2001, FAO <www.fao.org>
H Friedmann, The origins of third world food dependence in The Food Question. Profit versus people? eds H Bernstein, B Crow, M Mackintosh & C Martin, Earthscan, London, 1990
<www.oneworld.org/ips2/jul98/23_13_097.html>
B Crow, Moving the lever: a new food aid imperialism in The Food Question. Profit versus people? op cit
G Tansey & T Worsley, The Food System, Earthscan, London, 1995
C Hawkes & J Webster, Too much and too little? Debates on surplus food redistribution. Sustain, 2000 <www.sustainweb.org>
US food aid
United States Department of Agriculture s Foreign Agricultural Service <www.fas.usda.gov/food-aid.html>
Top ten recipients of WFP aid
World Food Program <www.wfp.org/aboutwfp/facts/2000.html>
Margin note
<www.fns.usda.gov/pd/annual.htm>

Part 2: Farming

32 — 33 MECHANIZATION
All data from: Production Yearbook, FAO, various editions

34 — 35 ANIMAL FEED
Grain fed to animals
World Resources 1996—97: A Guide to the Global Environment, Oxford University Press 1998 Table 10.3
The cost of producing meat
Joni Seager, The State of the Environment Atlas, Penguin, London & New York, 1995 p103
Peter Cox, Why You Don t Need Meat, Thorsons, London & New York, 1986
The grain drain
Peter Cox, Why You Don t Need Meat, Thorsons, London

& New York, 1986
Margin note
Greenpeace International <www.greenpeace.org/~geneng/reports/gmo/gmo019.htm>

36 — 37 MAD COW DISEASE
Meat and bone meal exports
<w3.aces.uiuc.edu/AnSci/BSE/BSE_MBM_Exporting_Map.htm>
Symbol: <www.oie.int/eng/info/en_esbmonde.htm>
Variant Creuzfeld-Jakob Disease
<www.cjd.ed.ac.uk/figures.htm>
BSE in the UK
<www.defra.gov.uk/animalh/bse/index.html>
BSE outside the UK
<www.oie.int/eng/info/en_esbmonde.htm>

38 — 39 INDUSTRIAL FARMING
Compassion in World Farming <ciwf.co.uk>
Chicken-meat production, Pig farming, Pig production
FAO database: < http://apps.fao.org/> November 2001

40 — 41 AGRICULTURAL R&D
Extent of public funding
Agricultural Science and Technology Indicators (Asti) <www.asti.cgiar.org>
Tropical–Temperate split, Public–Private split
Philip G Pardey and Nienke M Beintema, Slow Magic: Agricultural R&D A Century After Mendel, International Food Policy Research Institute <www.ifpri.org/pubs/pubs.htm#fpr>

42 — 43 GENETIC MODIFICATION
Genetic Engineering, Genewatch UK, Briefings: no10 (April 2000), no 13 (January 2001),
<www.genewatch.org>
Keeping control
Hugh Warwick, Syngenta — Switching off farmers rights Genetics Forum, ActionAid, Genewatch UK, Berne Declaration and the Swedish Society for Nature Conservation, October 2000
Super Salmon
National Geographic, May 2002, pp6—7
Golden Rice
<www.syngenta.com>
<www.purefood.org/corp/gericetoofar.cfm>
<www.greenpeace.org>

44 — 45 GENETICALLY MODIFIED CROPS
Commercial Cultivation of GM Crops, GM Crops, GM traits
Genetic Engineering: A Review of Developments in 2001, Genewatch UK Briefing no 17, February 2002 <www.genewatch.org>

46 — 47 PESTICIDES
Pesticide use
World Resources 2000—2001, Table AF.2

Types of pesticide sold
Agrow s Top 25, 2001 Edition, Arthur Dewar
Pesticide sales worldwide
Wood Mackenzie & Co, 2001
Regional sales
Agrow s Top 25, 2001 Edition, Arthur Dewar
 Globalization Inc. Concentration in Corporate Power:
 the Unmentioned Agenda , ETC Group Communique,
 Issue no. 71, July/August 2001 <www.etcgroup.org>

48 — 49 WORKING THE LAND
D Gow, Reshaping the landscape: Bigger farms but poor
 farmers , The Guardian, London, July 14, 2000, p28
Sustainable Agriculture in a Globalized Economy,
 Geneva, 2000, Section 4 <www.ilo.org>
**Agricultural workers, Declining importance, Increasing
 labor force**
FAO database < http://apps.fao.org/> April 2001
Wage levels
Agricultural wages in a number of countries: data in
 Sustainable Agriculture in a globalized economy, ILO
 2000 Table 10 <www.ilo.org>
The impact of Aids
State of Food and Agriculture, 2001, FAO, Rome
 <www.fao.org>

50 — 51 URBAN FARMING
<www.cnn.com/HEALTH/9802/06/hong.kong.flu/>
Food production in urban and suburban areas
Dongus, S, Urban Vegetable Production in Dar es Salaam
 (Tanzania) — GIS-supported Analysis of Spatial Changes
 from 1992 to 1999. APT-Reports 12, July 2001, p.
 100—144. Freiburg, Germany.
Smit, J.; A. Ratta & J. Nasr, Urban Agriculture: Food, Jobs
 and Sustainable Cities. United Nations Development
 Program, Publication Series for Habitat II, Volume One.
 UNDP, New York, USA, 1996
Stevenson, C.; P. Xavery & A. Wendeline (1996): Market
 Production of Fruits and Vegetables in the Peri-Urban
 Area of Dar es Salaam. Urban Vegetable Promotion
 Project, Dar es Salaam, Tanzania (unpublished).
The following chapters from Growing Cities, Growing
 Food — Urban Agriculture on the Policy Agenda,
 Deutsche Stiftung f r internationale Entwicklung (DSE),
 2000, Feldafing, Germany:
 Novo, M.G. & C. Murphy Urban agriculture in the city
 of Havana: A popular response to crisis ; Jacobi, P.; J.
 Amend & S. Kiango: Urban Agriculture in Dar es
 Salaam: Providing an indispensable part of the diet ;
 Potutan, J.; W.H. Schnitzler; J.M. Arnado; L.G. Janubas
 & R.J. Holmer Urban Agriculture in Cagayan de Oro: A
 favourable response of City Government and NGOs ;
 Garnett, T.: Urban Agriculture in London: Rethinking
 our food economy ; Yoveva, A.; B. Gocheva; G.
 Voykova; B. Borissov & A. Spassov: Sofia: Urban
 Agriculture in an economy in transition ; Mbaye, A. & P.
 Moustier: Market-oriented Urban Agricultural
 Production in Dakar ; Mbiba, B. Urban agriculture in

Harare: Between suspicion and repression ; Foeken, D.
 & A.M. Mwangi: Increasing food security through
 urban farming in Nairobi ; Torres Lima, P.; L.M.
 Rodriguez Sanchez & B.I. Garcia Mexico City: The
 Integration of Urban Agriculture to contain Urban
 Sprawl ; Kreinecker, P.: La Paz: Urban Agriculture in
 harsh ecological conditions ; Purnomohadi, N.: Jakarta:
  Urban Agriculture as an alternative strategy to face the
 economic crisis ; Yi-Zhang, C. & Zhangen, Z.
  Shanghai: Trends towards specialised and capital-
 intensive Urban Agriculture
Mougeot, L., Urban Agriculture: Definition, Presence,
 Potentials and Risks, and Policy Challenges . Paper
 submitted for presentation at the International Workshop
 on Growing Cities, Growing Food — Urban Agriculture
 on the Policy Agenda, Havana, Cuba, October 11—15,
 1999. Draft October 1999. Ottawa, Canada.
Smit, J, Urban Agriculture, Progress and Prospect:
 1975—2005. Urban Agriculture Network. IDRC, 1996
 Ottawa, Canada.
RUAF Fact Sheet on Urban Agriculture, food security
 and nutrition , Urban Agriculture Magazine, Special
 Issue for the World Food Summit, November 2001, p
 6—7. Resource Centre for Urban Agriculture (RUAF).
 Leusden, The Netherlands.
RUAF, Cuba — Ciudad de la Habana , Urban Agriculture
 Magazine, Special Issue for the World Food Summit,
 November 2001, p. 10. Resource Centre for Urban
 Agriculture (RUAF). Leusden, The Netherlands.
Moldakov, O.,   The Urban Farmers of St Petersburg
 Urban Agriculture Magazine 1, June 2000, p 24—26.
 Resource Centre for Urban Agriculture (RUAF). Leusden,
 The Netherlands.
Moustier, P., Urban and Peri-Urban Agriculture in West
 and Central Africa: An Overview . Paper prepared for
 SIUPA stakeholder meeting and strategic workshop, Sub-
 Saharan region, 1—4 November 2000. Nairobi, Kenya.
Castro, G. (2002): Cr a de especies animales productivas
 en zonas urbanas y periurbanas de le ciudad de
 Montevideo   (unpublished). Montevideo, Uruguay.
Mougeot, L., Urban Food Production: Evolution, Official
 Support and Significance (with special reference to
 Africa). Published by City Farmer. Vancouver, Canada,
 1994 <www.cityfarmer.org/lucTOC26.html>
Growing cities
Howson, Fineberg & Bloom, 1998
Dar-es-Salaam
Dongus, S. Urban Vegetable Production in Dar es
 Salaam (Tanzania) — GIS-supported Analysis of Spatial
 Changes from 1992 to 1999 in APT-Reports 12, July
 2001, p100—144. Freiburg, Germany.
Stevenson, C.; P. Xavery & A. Wendeline (1996): Market
 Production of Fruits and Vegetables in the Peri-Urban
 Area of Dar es Salaam. Urban Vegetable Promotion
 Project, Dar es Salaam, Tanzania (unpublished).
Russia: Global Environment Outlook-3, UNEP 2002
Margin note
<www.cf.ac.uk/claws/events/lang/sld065.htm>

52 — 53 FISHING AND AQUACULTURE

The State of World Fisheries and Aquaculture 2000, FAO <www.fao.org>

Fish catch

World fisheries production by capture and aquaculture by country (1999) <ftp://ftp.fao.org/fi/stat/summ_99/aqua_a-0.pdf>

The State of World Fisheries and Aquaculture 2000, FAO, Part 1, Table 2; Part 3, Figure 38 <www.fao.org>

State of fish stocks

The State of World Fisheries and Aquaculture 2000, FAO <www.fao.org> Part 3, Figure 37 <www.fao.org>

Changing balance of fish production

FAO Yearbook of Fishery Statistics 1999 — Summary Tables <www.fao.org/fi/statist/summtab/default.asp>

Where does fish come from? Where does it go to?

The State of World Fisheries and Aquaculture 2000, FAO <www.fao.org> Part 1, Table 1

54 — 55 AGRICULTURAL BIODIVERSITY

Genetic diversity

Jane Rissler & Margaret Mellon, The Ecological Risks of Engineered Crops, The MIT Press, Cambridge, Mass and London, England, 1996. Map based on research by Professor Jack R Harlan, University of Illinois, cited in National Geographic, April 1991

Vegetable and fruit varieties, China, Wheat varieties

Rissler & Mellon op cit

Endangered domestic breeds

World Watch List for domestic animal diversity, FAO, Rome, October 2000

56 — 57 SUSTAINABLE AGRICULTURE

Hungary figure: FAS online, Organic Perspectives, March 2002

<www.organic-europe.net/europe_eu/statistics.asp>

Increase in organic land in USA
Increase in organic land in EU

Minou Yussefi & Helga Willer, Organic Agriculture Worldwide 2002 , Stiftung Okologie & Landbau <www.soel.de/oekolandbau/weltweit.html>

Laying hens

Agricultural Outlook, April 2000, ERS/USDA <www.fas.usda.gov/htp/organics/organics.html>

Part 3: Trade

60 — 61 TRADE FLOWS

All data: World Trade Organisation <www.wto.org>

62 — 63 ANIMAL TRANSPORT WORLDWIDE

All data: FAO database <www.apps/fao/org.uk/>

64 — 65 ANIMAL TRANSPORT IN EUROPE

Animal Transport within Europe

Peter Stevenson, Live Exports, Compassion in World Farming, 2000 <ciwf.org.uk>

Foot and Mouth Disease

<news.bbc.co.uk/hi/english/uk/newsid_1334000/1334466.stm>
<www.guardian.co.uk>

66 — 67 FOOD MILES

Milk miles

FAO 2002 <www.fao.org>

Pollution

W J Dijkstra and J M W Dings, Specific energy consumption and emissions of freight transport. Centre for energy conservation and environmental technology, Delft, The Netherlands, 1997

Air Freight

Calculatations by Andy Jones, Eating Oil, Sustain, London 2002. Based on data sourced from UKROFS and a survey of supermarket stores June — August 2001; distance tables for air miles <www.indo.com/cgi-bin/dist>; environmental impact of airfreight in Guidelines for Company Reporting on Greenhouse Gas Emissions, Department of the Environment, Transport and the Regions, London, March 2001.

68 — 69 SUBSIDIES AND TARIFFS

Agricultural Policies in OECD Countries: monitoring and evaluation 2001 highlights. OECD, Paris

World Trade Organization: Member s Usage of Domestic Support Categories, Export Subsidies and Export Credits <www.wto.org/english/tratop_e/agric_e/ngs12_e.doc>

Make Trade Fair <www.maketradefair.com>

Tariffs

World Trade Organization <www.wto.org/english/tratop_e/agric_e/negs_bkgrnd07_access_e.htm#thetariffs>

Support to Producers

OECD Database:Table III.3 and Table II.1 <www.oecd.org/oecd/pages/home/displaygeneral/0,3380,EN-statistics-131-4-no-no-no-131,FF.html>

OECD subsidies post-GATT

OECD Database Table II.1

Value of subsidy

Agricultural Policies in OECD Countries: monitoring and evaluation 2001 highlights. OECD, Tables 1.6 & 1.7

Agricultural support

USDA, cited in Financial Times, London, July 23, 2001

UK direct payments to agriculture

Agriculture in the United Kingdom 2001, Department of Food, Agriculture and Rural Affairs (DEFRA) London, 2002 <www.defra.gov.uk>

70 — 71 TRADE DISPUTES

<docsonline.wto/org>
<www.naftaclaims.com/>
<www.dfait-maeci.gc.ca/tna-nac/NAFTA-e.asp#Notices/>

Going Bananas

Anne-Claire Chambron Straightening the bent world of the banana, EFTA, February 2000 <www.bananalink.org.uk/>

<www.fao.org/waicent/faoinfo/economic/ESC/esce/escr/bananas/banmie.htm#BIN>

72 — 73 DEVELOPING TRADE
Food Production and Trade
FAO, 2002
Tumbling cocoa price
UNCTAD Commodity Price Bulletin
More Trade, Less Food?
International trade statistics 2001. World Trade
  Organisation, Geneva, Switzerland.
  <www.wto.org/english/res_e/statis_e/its2001_e/stats2001
  _e.pdf>
Food Balance Sheets. Food and Agriculture Organisation,
  Rome Italy. <www.fao.org>

74 — 75 FAIR TRADE
Fairtrade Labelling Organisation (FLO)

PART 4: Processing, Retailing and Consumption

78 — 79 STAPLE FOODS
<www.fao.org/inpho/vlibrary/u8480e/U8480E07.htm>
FAO database, Nutrition, Food Balance Sheets:
  <http://apps.fao.org/>

80 — 82 CHANGING DIETS
<www.fao.org/inpho/vlibrary/u8480e/U8480E07.htm>
FAO database, Nutrition, Food Balance Sheets:
  <http://apps.fao.org/>

82 — 83 PROCESSING GIANTS
Top Ten Global Food Companies
Company reports, 2000
Top brands
Interbrand/Citibank, cited in R Tomkins, Coca-Cola loses
  its fizz , Financial Times, July 19, 2000, p17
Concentration
Henrickson-Heffernan Concentration of Agricultural
  Markets Department of Rural Sociology, University of
  Missouri, Columbia, USA

84 — 85 RETAIL POWER
Consumer Expenditure for Farm Food 2000, March 2002
  <www.fmi.org/facts_figs/keyfacts/stores.htm>
US Food Marketing Institute <www.fmi.org >
European ECR: <www.fmi.org/supply/ECR/ May 3 2002>
Turnover
<www.fmi.org/facts_figs/keyfacts/weeklysales.htm>
Floor space
<www.fmi.org/facts_figs/keyfacts/storesize.htm>
Intensity of sales
Author s calculation based on:

<www.fmi.org/facts_figs/keyfacts/weeklysales.htm>
<www.fmi.org/facts_figs/keyfacts/storesize.htm>
Top retailers in USA
Ali Urbanski, The Slow Shakeout, Progressive Grocer
  <www.grocerynetwork.com/progressivegrocer/images/pdf/t
  op50.pdf>
Top retailers in UK
IGD Press Release May 7, 2002
Top retailers in Europe
IGD Press Release May 7, 2002
European retailers Go East
Euromonitor/companies/food issues/retail/FT eastward
  drive 07 01
Wal-mart takes over
Financial Times, London, March 15, 2002, p21
Financial Times, London, March 17, 2002, Lex column

86 — 87 FUNCTIONAL FOODS
M Heasman and J Mellentin, The Functional Foods
  Revolution: Health People, Healthy Profits?, Earthscan,
  London, 2001

88 — 89 ORGANIC FOOD
Minou Yussefi & Helga Willer, Organic Agriculture
  Worldwide 2002 , Stiftung Okologie & Landbau
  <www.soel.de/oekolandbau/weltweit.html>
Andy Jones, Eating Oil, Sustain, London 2002
Agricultural Outlook, April 2000, ERS/USDA
<www.fas.usda.gov/scriptsw/AttacheRep/gain_display_rep
  ort.asp?Rep_ID=135683121>
Sales of organic food
International Trade Centre (UNCTAD/WTO (ITC),
  <http://www.intracen.org/mds/sectors/organic/overview.p
  df>
 Organic Agriculture Worldwide, op cit, p13
Purchasing organic produce in Europe
 Organic Agriculture Worldwide, op cit p86
Purchasing organic produce in USA
Rudy Kortbech-Olesen and Tim Larsen, International Trade
  Centre UNCTAD/WTO, 2001
  <www.intracen.org/mds/sectors/organic/welcome.htm>
Consumer spending on organic food
Organic Agriculture Worldwide, op cit. p87

90 — 91 FOOD ADDITIVES
Leatherhead Food RA
  <www.lfra.co.uk/lfra/lfra/press736.html>
E Millstone, Additives: a guide for everyone, Penguin
  Books, London 1988
Global market for sweeteners
  <scup.sric.sri.com/Enframe/Report.html?report=ANTIO0
  00&show=Abstract.html>

92 — 93 EATING OUT
All data: Foodservice Intelligence Ltd

94 – 95 FAST FOOD

P Fieldhouse, *Food and Nutrition: Customs and culture.*
  Chapman & Hall, London, 1995, p.210
The Guardian G2, page 2, London, May 2, 2002
Jeffrey P Koplon and William H Dietz, "Caloric Imbalance
  and Public Helth Policy", *Journal of the American
  Medical Society*, October 27, 1999;
"Resident Population Projections, by Age and Sex",
  Statistical Abstract, p17, quoted by Eric Schlosser in *Fast
  Food Nation*, Houghton Mifflin, New York, 2001,
  Penguin, London 2002
McDonalds worldwide
<www.mcdonalds.com/corporate/investor/financialinfo/fin
  ancials/media/downloadablefinancials.xls>
BurgerKing
Supplied by BurgerKing
KFK and PizzaHut
<www.triconglobal.com/investors/units.htm>
Microwaves
<www.sharp.co.jp/sc/gaiyou/news-e/000204.html>
Ready rise
<www.superpanel.com >

96 – 97 ALCOHOL CONSUMPTION

Christopher J L Murray and Alan D Lopez, *The Global
  Burden of Disease*, WHO/Harvard School of Public
  Health/World Bank, 1996
**Wine consumption, Beer consumption, Spirits
consumption**
World Drinks Trends, Productschap voor Gedistilleerde
  Dranken/World Advertising Research Center Ltd, Henley
  on Thames, 2002
Regional consumption
Global Status Report on Alcohol, Substance Abuse
  Department, World Health Organisation, Geneva, 1999

98 – 99 ADVERTISING

Money spent on food advertising, Coca-cola advertising
*World Advertising Research Centre Marketing Pocket
  Book 2001*, Global AdAge <www.adage.com>, Zenith
  Media and Marketfacts 2001
Currency conversions from published figures and from
  <www.xe.com>
Population estimates for 1999 from *WHO World Health
  Report 2000*
Big Spenders
Global AdAge, Top 100 Marketers (2002)

100 – 101 CITIZENS BITE BACK

Consumers International
  <www.consumersinternational.org/members/index.html>
International Union (IUF)
  <www.iuf.org.uk/en/affiliates.shtml>
**International Baby Food Action Network
  (IBFAN)**<www.ibfan.org/english/gateenglish.html>

Photo credits

INDEX